THE CARBON CONNECTION

THE CARBON CONNECTION

By
LEONARD RIDZON
who lived and relived this book

AND

CHARLES WALTERS JR.
who wrote it all down

ACRES U.S.A.
KANSAS CITY, MISSOURI

THE CARBON CONNECTION

Acres U.S.A.
P.O. Box 9547, Kansas City, Missouri 64133

ISBN 0-911311-24-6
Library of Congress catalog card number: 90-061305

To Barbara Ridzon,
who performs miracles

CONTENTS

A Note From The Publisher

Leonard Ridzon had a vision one day some ten years ago. Suppose, he said, we could unlock the biological energy contained in coal the way we have released its mechanical energy with the agency of heat, then these stored reserves of carbon could furbish and refurbish agriculture for all the centuries between now and the next Ice Age. Ridzon was and remains a full time farmer. He realized then—and he realizes now—that global commodity prices made it impossible for a farmer to stay solvent using salt fertilizers and toxic rescue chemistry. However, two centuries of intensive farming in most parts of the nation had done violence to the carbon in the soil.

Coal, carbon and the carbon connections became a consuming interest to Leonard Ridzon. But there was a paucity of information on the subject, especially soil carbon. The literature yielded little until farmer-scientist Leonard Ridzon discovered Siegfried Marian and the thousands of experiments he conducted in his laboratory in pursuit of the carbon connection. Marian also dreamed of unlocking the secrets of coal, and he suggested that the problem be faced with all deliberate speed. Marian lived and worked in post-war England as something of an outsider, for the world of agronomy had progressed no further than the reckless conceits of Justus von Liebig. His death at an early age cut short a run of progress that has still to be fully appreciated. Marian often pointed out that other disciplines had made progress. Engineering, manufactures, commerce, scientific pursuits of every ilk, all had taken form, gathered speed and moved ahead, decade after decade, but not agriculture. It languished because it had become lock-stepped into a faulted system of thought and action. Those who sought to communicate sound biological principles were downgraded and subjected to the sneers of academia. Marian endured—and then he passed away at an early age, largely unknown. Before he died, he made a record of his findings. What he wrote down with a pen cannot be removed with a hatchet.

Leonard Ridzon was hard at the task of unlocking the carbon secrets of coal when he rediscovered the works of Marian. Ridzon had already made his results a matter of record. Now he plugged in many of the reasons *why*!

The Carbon Connection is a book for today's literate eco-farmers. It may well help recapture the values lost with the public policy drive to eliminate the family farm. It seems to remind America that the family farm has been and remains the most valid and vital of its teaching institutions. It is therefore ironic that while others gain fame for foisting on the world destructive ideas, Leonard Ridzon is about the business of healing the good earth naturally, generally unrecognized and somewhat unknown. This is no tragedy. For it is the lesson that counts, and with this little book Leonard Ridzon presents a very valid lesson.

Foreword

Before I start my report on the subject of carbon and coal, let me tell you a story. It is a paraphrase of what was set down in words like these by a young scientist.

This scientist had met Papa Sandor when the home he lived in was located in Hungary. It was 1917. Not long after that an evil man named Wilson—some called him *der schlechte* Wilson—had gone to Versailles and changed all that. The several leaders cut up the old Austrian-Hungarian state into small nations and altered many things not in the best interest of humanity. Those small states enjoyed only a few years of independence before falling to Hitler, and after Hitler's defeat to Stalin. So Papa Sandor ended up living in Czechoslovakia, even though he didn't move. He knew only Hungarian as a principal language, and after that, a little German. He'd picked up his German in the Austrian army.

The young scientist hadn't seen Sandor in over a decade. Now he wanted to see him again. As so often happens, life and education prompted him to recall something Papa Sandor had hinted at over ten years ago, a secret that made the old peasant's land and produce superior to all the neighbors. That is why the young scientist decided to go via Hungary to see a Hungarian now living in Czechoslovakia, even though he had never moved.

The scientist made a two stage journey. First he had to travel to Budapest. So on a Sunday—after a rugby match between Hungary and Austria—this young scientist took the Orient Express on the London-Istanbul run, arriving in Budapest at tea time. He visited old associates, and a day or so later he took the night train into Slovakia, stepping onto the platform the next morning in a town called Kassa when it belonged to Hungary. It was now called Kosice. The politicians had moved boundaries, but they could not move the Carpathian mountains, which were splendid under a cover of sunlit snow. It was as if God's own nature had confronted the politicians with a challenge: *Who asked you to speak for the land?* Papa Sandor was much more capable of speaking for the wind and the soil, the north, the south, the east and the west.

To get to Papa Sandor's place, the young scientist had to hire a horse-drawn sleigh and travel for three hours. Sandor was there, but he was very old, and the young scientist wondered whether he would take his secret of the soil to his grave with him. Still, there was a sparkle in his eyes. It suggested a little deviltry and a great lust for life. The young scientist first thought that Papa Sandor would not recognize him, but this uncertainty faded quickly, and the old man remembered—and he quickly got into the business of recollecting everything that had happened since they had met some ten years ago. It did not come to Papa Sandor's mind that the young scientist had come to learn his secret. He thought the younger man had simply come to visit him, and he was happy about that.

The young scientist let it go at that. He decided he would wait for the right moment, and then he would bring up the matter of the farmer's preparation, based on sawdust. This was due to happen after a meal, and so it did after they had enjoyed hot soup and dumplings, heaps of fat, salty bacon covered with a crust of paprika and dried prunes, which were eaten raw. Out came a bottle of wine, and then Sandor was ready for the story the young scientist had heard ten years ago.

This gave the young scientist an opportunity to become inquisitive. Could he tell him more about it, and how he was making it? The world may be thankful to him, but he became at once suspicious. "Ah, you clever people studying at universities and whatnot should know better than an old peasant who cannot read and write. Why don't you find it out yourself, and make much better stuff than I can, ah? You do make plenty of stuff not worth to be given to pigs, and you call it medicines, or don't you? Ah, never touch these rubbish things, they help nobody but the chemist to get rich and fat. I am not feeding chemists. Better to feed my pigs."

He said much more, but he changed his tone gradually, and said: "Ah, you can have a bottle of the stuff. Take it home, analyze it or whatnot, and you will know all about it. I am not going to tell anybody how I am making it. If you clever people are not clever enough to find such a simple thing, you are not worth walking about on this earth. You know how to eat what the earth gives you, but you do not bother to know how the earth does it, or do you? Look how they now dust the earth with some dirty and sharp stuff. I bet they have no idea what this stuff does and why they put it on. Now, have another glass of wine, is it not lovely? Ah, the earth which grows this stuff is clean earth, clean, do you know what clean earth is? Bet you have no idea. Well then, clean earth makes clean crops, and clean crops make clean men. Those sly devils who unload stinking stuff on the earth make it dirty, and then they get dirty crops. Ha, they shake their heads when they find pests eating their fruit. And the pests laugh too. These pests are clever things. Give them a bit of soil, and in they go. They make their homes in soil and get busy making more and more little pests, without marrying."

Here the young scientist had to interrupt the stream of Sandor's words by asking: "Do you never use raw manure from your animals?" His reply was: "Me? Using this dirt on my good earth? Making homes for pests? How can you ask such a silly question?"

"But you are keeping animals, so what are you doing with the manure?" He had expected this question because his answer was rapid. "Ah, they [meaning other farmers] come and fetch it. They clean my stables and take it away. They pay me for it."

When the young scientist uttered words of misgivings about such behavior which was contrary to all accepted techniques, Sandor said: "If I were a young man, you could tell me that I am doing wrong, but I am a very old man, and I have lived on this farm since I was born. It is a long time, you know, and I have the best crops, the best animals, and I have money in the bank, no debts. No, I am going to teach you a lesson. You are young, and one day the words of old Sandor will be found to be correct. Well, look through this window, what do you see? Forest, and nothing else on all the hills. Where this house stands there were trees standing. The earth is full of waste from these trees, and it will be full another few thousand years. Waste is waste and must go. Before the waste has not gone, the soil is unclean. You would not keep waste accumulating in your place? Can you tell me why we plow and dig? I can tell you. Not to break up the soil, but to get the waste to the air so that a little of it shall be burnt by the air. Every little bit of waste burning away increases the yield of the earth by that little, and the more we plow, the longer we plow, the better will the earth become. It is silly to put more waste on old waste, and so I am not putting dirt on my soil."

The scientist decided not to carry the argument with Sandor any further. He was talking with a fire of conviction which sounded true, and there was no doubt that his method was very successful. He therefore said: "Well, Sandor, you are probably right in your case, I can see it clearly." But Sandor had another lot of words, and he threw in an argument worth recording. What was first, animals or trees and grass? Have animals made the forest and grass, or have trees and grass made the animals? If animals can make plants grow, why were they not first? The first plants came without animals, and all further plants can live without animals. Plants come from seeds of plants and from clean earth, like the first trees and the first grass came from clean earth."

So the two talked for another hour or longer before the scientist was led into the bedroom where a nice log fire was burning, and a candle was burning in addition to the petroleum light hanging down from the ceiling. The next morning Sandor presented him with a bottle of his "medicine," which he carefully packed into his case, and the young scientist started his return journey after lunch.

The young scientist was Siegfried Marian. And what he learned from a bottle of Papa Sandor's stuff both backgrounds and illuminates everything we know about the carbon connection.

1
Carbon

Organic has always been a troublesome word, not only for the practitioners of ecological agriculture, but for the hard chemistry boys as well. The reason is that carbon is not well understood on the street, and one can be allowed to wonder whether the people in the ivy towers have a firm grasp on the subject. Yet most of chemistry is built on a *foundation* of carbon. *Foundation* may not be a perfect term, but it will do until the grammar of this subject can be fleshed out in the pages and chapters that make up this book.

Carbon is the grand combiner, if that term can be permitted. It combines with all other elements to form compounds so numerous that in theory the full roster cannot be described. It is for this reason that carbon chemistry has been sequestered from the rest of chemistry and called *organic chemistry.* Most carbon compounds are parts of living matter, or they have been formed by living matter. The science of carbon dating is based entirely on the proposition that the substance being dated was once alive.

For now it is enough to say that all living matter is constructed on a frame of carbon, chemical symbol C, atomic weight 12.011. Picture, if you will, the skeleton of a building, with "I" beams bolted and riveted together so that the eerie windows can be filled in with the other building materials needed to complete the structure. Without the frame the structure will not hold together. Now transport the general idea to the plant building business.

Plants in fact synthesize this carbon frame and then fill it up—according to a genetic program—with carbohydrates, fats, proteins, whatever. In turn this carbon frame and its building materials are consumed by animals and human beings to form the frame of a new existence. Simply stated, we are governed by how plants build their frames and fill them up to confer on both animals and human beings an essential variety of compounds. This common denominator tells us that human diseases are due to a lack of certain carbon compounds.

But the carbon frame comes first. Moisture and energy from sunlight cooperate to turn this carbon into a gas—the ubiquitous carbon dioxide that is loaded into the atmosphere. That is why we say that the carbon frame must be built from gas, a liquid and energy. More specifically, sunlight serves as an on-scene spark plug to energize both the gas and the liquid so that the carbon frame can take form, gather speed and grow to fruition.

It isn't that simple, of course. There has to be a source of energy to pump the moisture from the soil into above-soil parts of the plants, a pump that defeats the law of gravity with its own law of levity, so to speak. This energy is generated by the battery effect between soil constituents and mild acids exuded from the roots of plants.

Farmers often wonder aloud why the moisture supply fails. In the coffee shops and hardware stores, they grouse about drought and send out excuse feelers to their neighbors, always looking for a reason for failure. Few pause to reflect on the carbon connection and its role in fine-tuning the above-ground level of energy called sunlight to the underground energy that services nature's own pump. Only under conditions of balance can the plant's demand for moisture be met, a prerequisite for proper construction of the carbon frame.

It is in this underground laboratory for energy formation that carbon plays the most vital part. Most of the carbon frame which life builds, and on which it builds, returns to the soil as organic matter so that newer carbon frames can be built from that gas called carbon dioxide, water and sunlight energy.

Carbon must be converted from one phase to another before it can give us permission for life. Unfortunately the rhythms of the universe have set up fault lines and traps. Plant growth has not always yielded gaseous carbon dioxide. Freshmen students in geology can recite the many systems the Creator has used to store energy as coal, oil and several shades of humates, Leonardite and energy packed shale. Here we will pause only to note that most solid carbon compounds are not available for the promotion of life. Only the gaseous state called carbon dioxide serves the fresh synthesis of life operating with the agency of sunlight, and only a part of the solid carbon state operating with the agency of water—wherein chemicals reactions with soil particles are possible—can serve the pump requirements of crop production.

Not many farmers—or professors, for that matter—know that there is more carbon in cultivated soils and under forest canopies than there is in all the coal mines of the world. Unfortunately, very little of this carbon is of the right character. It is inert and too often defies the farmer's trillions of unpaid workers, the microorganisms, a successful go at breaking the fault. Whether the microorganisms can break the fault or not via digestion depends on the inert character of the carbon.

It would be correct to describe as *constipated* the digestion process, especially when high concentrations of metals or hydrogen are present—soil sickness, if you will. In alkali soils, *soil diarrhea* would be a better term. This is simply to say that soil conditions determine whether carbon contained in organic matter can be considered of nutritional value. It must be digested by the microorganisms before a complex of carbon compounds can release mineral elements vital to the plant growth.

It is this freedom of conversion that is at issue, and yet so little understood. The soil is basically made up of 50% carbon in different compounds. We have carbon in an hexagonal form. Diamonds are carbon representing the hardest and purest form. The next hardest form, moving down the scale, is the carbon in anthracite coal, and after that comes bituminous coal, then lignite. The carbon in Leonardite comes next, then muck, humus, the carbon in plants, and finally carbon dioxide, carbon monoxide and hydrocarbons in the soil and air. All of these are in the environment in varying degrees. They all have their degrees of availability, depending on the age of the structure, the heat, the compression, the pressure, the degree of gasification. All of these factor into the equation of life. Carbon is where and when life begins. It is the womb of life, so to speak. Every one of the carbons has a relationship to nature. Each has a job to do. In hydrocarbons, it is a method for moving energy. In a diamond, it is locked up and it can't be used by anything, except perhaps as a reserve of economic wealth.

Add a word to your vocabulary—activated carbon. To make activated carbon they take a unit of carbon, heat it and cause it to expand under controlled conditions until it starts to fracture or crack. This creates an increase in the areas for absorption of toxins. The concept of activated carbon is to provide a surface on the inside equal to the surface of the outside of a unit of carbon. The inner surface can only be penetrated through the fracture, and the degree of fracture determines the energy field that can get into the inner chamber. The numbers 22 to 532 suggest that there is a wide range in the field of activation. The average person knows very little about activated charcoal.

Activated charcoal is used as a purifier. It is used to remove acid from the stomach in humans. All water filters use activated charcoal. And that's what it does in the soil. Activated carbon is activated to absorb specific energy fields. However, the carbon in the soil is a different type of carbon. Originally it was in perfect balance and could serve the purpose of purifying. The carbon that is in NutriCarb coal has both positive and negative energy fields in equal amounts, making it the perfect purifier. This is the reason why NutriCarb is working so well in the area of purification. A method of testing I have developed allows me to test all coal to determine how much energy the carbon can or will absorb. Many coals exhibit toxicity in varying degrees. This causes them to annihilate microorganisms and not to break down or absorb energy causing it to be toxic. Coal carbon can test toxic in either negative or positive terms. When this storage capacity is used up and the carbon can hold no more toxins, new carbon must be added or the bacteria will be forced to use the old carbon, thus releasing unwanted toxins. This causes unwanted weeds and grasses to grow and unwanted nutrients to debilitate farm crops. When an energy field in the carbon is filled, and it can hold no more, then this energy becomes part of the water in the ground. Nitrates and herbicides in underground water suggest a carbon structure with its absorbing capacity saturated.

All energy is electrical in nature. It must search out opposites to attach to or equals (same fields) to join, much as drops of water become a creek, a river, an ocean. Water drops are an energy field, a storing, moving energy unit.

So, what we have in carbon is a real natural phenomenon. Carbon can attract all energy fields, or electrical impulses, and store them in the carbon units if they are in sufficient concentration to be toxic to the environment. The bottom line is that carbon's purpose is number one, to act as a purifier. We are all familiar with the charcoal filters that we put on the water source to purify the water. This is exactly what it does to the soil. In the case of water purification, we put a little charcoal filter on to the water flow, turn the water on. The water goes in polluted and comes out pure. It is instantaneous. The carbon in the soil functions exactly the same way. As the toxins reach the carbon structure, the carbon sucks them up right away, like a sponge.

A toxin is anything in nature that is present in a quantity higher than is needed to sustain life, thereby destroying life. These are represented by high atom numbers per molecule. The heavy metals, being pure energy, are attached by bacteria to the surrounding rock formation. Let's look at an acid, a pure ener-

WHAT IS LIFE

Man is too sure of himself. Mastery of the ground on which we live is by no means accomplished, although we live and behave as if it would be a fact. We still have not solved the problem of what is life. Many times have we seemingly come quite near to a formula or a discovery which looked like a solution. In the beginnings of human civilization man thought that by the discovery of fire he had solved the mystery of life. With this mystery he was to win mastery. He set fire to the forests in which his enemies, the wild animals lived, and hoped that with fire at his disposal he had survived for ever.

It is indeed amazing how we still exist on modifications of our first great discovery. Fire, heat, we convert into other forms of energy, and the mastery of energy we call power. In the beginning we relied on trees which grow out of our soil. We invented the axe—the second greatest discovery—and won energy (power) from other living things, like trees. Then we discovered by digging underground, that there was "fire" hidden, and we dug for it. The new material we call coal. Few of us look at this precious material with the right reverence. We assume that it is there and we have to get it. Are we really much concerned about how it has come to be there?— Siegfried Marian gy. By passing an acid through carbon, the carbon takes the acid out of the liquid, leaving the carrier behind. The carbon has purified it.

If you have no carbon in the soil, you have a desert. There is no carbon in the Sahara Desert. There is little available carbon in a living desert, such as our California and Arizona areas. In other words, it takes the carbon to hold the soil together or to flocculate it and hold it in the right structure. Carbon is to the soil much as mortar is to bricks in making a wall. As noted above, the carbon in a diamond is extremely hard and pure. Diamonds are used as cutting tools. In anthracite coal, the carbon generally represents 88 to 98% of the product. It would probably make a good material for water purification. This is quite different from bituminous coal or humates or any of the various steps along the way. Raw humates are raw energy and could be toxic in some forms. The carbon form with the most promise is called bituminous coal. After that there would be lignite, Leonardite, muck, humus and organic matter. In each case the carbon would be less able to attract because it exhibits less purity and because it has absorbed toxins in the environment, namely chemicals, herbicides, etc.

Siegfried Marian was able to determine that the carbon in the diamond is hexagonal crystal carbon. He found that there was in the unit of carbon an element he called "biogenic carbon," which represented a half of 1% of the total. Biogenic carbon is indestructible to any type of disintegration. He found that when the bacteria took a unit of carbon and reduced it by 99.5% to humus, the 1/2% release represented "biogenic carbon." He called this the spark of life because this carbon also appears to be the start of life. Without the biogenic carbon, no life could start or exist. This is the lesson that has been put on tap with NutriCarb. The carbon in NutriCarb converts at a ratio of one pound NutriCarb to 100 pounds of humus. Independent laboratory testing has revealed that with NutriCarb we have had an increase in the soil of 100 pounds of humus per acre. This, in fact, proves Marian's theory concerning coal. Coal holds vast amounts of biogenic carbon. The conversion of carbon—which in bituminous coal runs approximately 40 to 70% carbon—and its release would help to rebuild the nutritional building blocks in the soil first, then in the plant. Other coals fall below bituminous. Muck has a one to one ratio, or less. This ratio tells us that his theory is correct because we are getting this actual conversion and release of biogenic carbon today, forty years after he wrote his hard to find reports. If the coal carbon is toxic because of absorption or because it was never pure, then it will not work. The carbon in lignite would be more of a hydrocarbon than the carbon in NutriCarb or bituminous coal. Lignite coal carries about 80% to 85% carbon. This indicates that it would be a better coal for purification than bituminous coal. But in actuality we find that it doesn't have that value because it doesn't have the conversion ratio. If you put a pound of lignite in the soil, you will get about a one to ten conversion. There is a pecking order when it comes to carbon sources, as related above. Anthracite rests at the top. It is the most valuable. It is the least ready to go, albeit only to burn.

NutriCarb has not been able to develop a method for using anthracite coal. It doesn't work for agriculture because it is too hard. That is why the NutriCarb process has dropped down to bituminous. NutriCarb, basically, is bituminous coal. It is a blend of coals mined in Ohio which I have been able to identify as deposits that will work when blended together. The identification and blending process is proprietary. The next carbon level is lignite. As soon as you go from bituminous to lignite, you go from a crystal carbon to a hydrocarbon. Down the ladder is graphite, which appears to be a liquid. It is, however, nothing but a hydrocarbon in a semi-solid state. The next source is Leonardite. Leonardite is nothing but a higher form of humus, or a very low form of coal. It has very little energy value as far as burning is concerned. A lot of people are using Leonardite and claiming great things, and this may be true. It all depends on variables. We hear people having great results with chemotherapy and getting rid of cancer, but we hear an equal number not satisfied with it. There are a lot of variables and unknown factors, and generally we do not hear of the failures. We do hear of the successes. The humates are derived from the coal.

Humates can be found in all forms of coal. They are in anthracite, lignite, Leonardite, and in muck. Down the scale, below muck, we get to humus, peat, organic matter. This is the spectrum with which the farmer can work if he is tapping into carbon technology.

Coal, as the scientific community tells us, is ancient vegetation. Today we know that the amount of material that comes out of the soil is insignificant in terms of the total crop. Experts bandy figures of 2 to 8% out of the soil—so we can argue from now to eternity and never accomplish anything. Let's establish that the greatest amount of energy comes out of the air. Here we have fusion. The least amount of energy comes out of the soil—fission. Conversely, we now look at the soil the way we would look at a fire. You need a match to start a fire, but you also need small twigs that can be easily ignited. To build the fire bigger, you have to add larger pieces of wood. We see fission at this point. With our carbon connection we have the same thing. We have carbon in the soil which is converted to humus. It is rich in energy. This is then converted to a source of energy for the plant and the plant then becomes large and falls down—fusion. We find muck in the bottom of all the rivers and ponds. Where did the leaves fall? They fell on top of the pond. Where did the bugs that got in the middle of the pond that got on that leaf go? They all sank to the bottom. This is all part of the environmental cycle. We know that there is no air at the bottom of a lake, so the only breakdown that can occur takes place with the agency of anaerobic bacteria. Everything has its place in the life and death cycle.

When coal is recovered and processed for distribution to the fields, the connection with good plant production depends on microorganisms. Under proper conditions, bacteria in the soil break the carbon down to a usable energy source for plant life. This can be compared to starting that fire with twigs. If you can't get the plant out of the soil, it is not going to take in carbon dioxide from the air, or carbon monoxide and all the electrical energy fields that would be attached to that carbon structure. Recently, the Environmental Protection Agency came out with a finding. EPA now says that carcinogen molecules are attached to carbon dioxide in the air. These are toxins. When the plant takes in carbon dioxide, it also takes in corresponding carcinogens. This explains how pollution spewed into the air returns to haunt plant cells, finally human cells and protoplasm. The pollution modern industry puts into the air is mostly negative energy. We are talking about burning coal, about burning oil, gasoline, fuel oil, etc. Oil is a composite of more negative energy than positive energy. Emissions from chemical factories and plastic factories and automobiles all come from crude oil. We put these pollutants into the air through that mechanism of industrialization and now the plants are sucking them in together with the required carbon. It is the function of plants to deposit toxins outside of the life cycle. This is very much like a bullet in the body that is not removed. The body will attempt to seal it off for the protection of life.

Carbon at this point serves as the police force in the soil. It does not allow everything to enter the plant's structure. When carbon is failing, or is locked up, or there is no carbon, then all of the good and bad flock into the plant structure. Now the plant has to get rid of the bad elements, and if it takes too much energy to get rid of toxins or growth inhibitors, then the plant will no longer reproduce itself. It will become sick and may die, or at best produce misfits. A subclinically sick plant invites insects to take it apart. But when the carbon is the police force protecting the plant, the plant utilizes it to delete the misfit elements from the environment. If the police force allows the toxins to enter, then the plant has to get bad actors out of the air to grow.

The consequences suggest themselves. Look at hybrid seeds. These seeds are designed to cohabitate with negative energy. Carbon now plays only a small part in the growing of crops because it is effectively locked up or gone. Some fifty years ago most of the soil carried about 100,000 pounds of carbon in an acre of soil in the top 6.33 inches. Improper agronomy advice has served to diminish our carbon bank account, and what is left is so locked up it is not functioning. Also, as a consequence of changing the structure of plants, our scientists have made them insect prone. And since they are insect prone, their very existence is an insurance policy for the sale of lots of rescue chemistry, this to see if growers can save the crop from human folly. Insects are nature's above ground police force. Because they fly around, they have the greatest life in the world. Basically they eat and have sex. That's it. Now, if you ever want to come back again you want to come back as a bug and hope that they don't use pesticides. The bugs land on a plant, test it to see if this plant will provide the food it and its family need. If the insect finds that a plant will provide the food it needs, friends and associates soon arrive by the busload. On the other hand, if the plant does not have clinically ill food—the insect's preference—all it leaves is a hole in the leaf, and goes on to the next plant. If the bugs land on a plant and the plant is not sick enough to die but is not well enough to fight off the bugs by developing antibodies or chemicals to shun the little critters, then the bugs are going to attack. A healthy plant will fight off the insects. When the bugs become ill, they leave.

At this point, carbon has no role because the insects are doing above ground what the carbon did not do or could not do below the surface of the soil. The reason that a plant needs to have a complete cycle in the soil is because each time the plant goes into a growth sequence, it must get its start from the soil. It is the function of carbon to make sure that the materials that go into the plant are proper for growth and health. All reproductive life seeds are generated by energy out of the air. Finally, it is very difficult to get a proper intake of nutrients without the carbon complex in the soil. Thus our dilemma. Carbon is one of the least discussed topics when farmers and scientists talk about plant growth. Not only *least discussed,* but least researched. Other than Siegfried Marian, few observers have even mentioned it as a matter of importance to crop production. Most agronomists believe that carbon in the soil has no effect on plant life. Understandably, these same people believe carbon in coal has no value for agriculture.

Few farm consultants even deal with the water crisis that toxic technology has accounted for, except to provide a little lip service. Yet it is a fact that water below pH 7 eats up pipes, concentrates heavy metals and otherwise imposes debilitating burdens on farm animals. I know of nothing that will clean up water in the calf's drinking cup like bone charcoal. This carbon form comes from cattle bones and has something called hydroxyapatite. This is the part of bones that attracts heavy metals. Bone charcoal is made by firing bones at 1,800 degrees Centigrade. The very pollutants that land in animal or human bones are picked up by bone charcoal, meaning chlorine, pesticides, insecticides, and—not least—toxic items such as aluminum, mercury, copper, fluoride and sulfur. Here as elsewhere the answer depends on making and understanding the carbon connection.

2
The Basis of Life

It is useless to study carbon without taking a good, hard look at nature. There we see all living things composed chiefly of carbon artfully combined with many other elements in configurations too numerous to inventory. Thus any look at nature has to start with a look at carbon. It is the open sesame to understanding the secrets of life.

There are, in fact, two basic carbon forms—the diamond and graphite—forms that are at once somewhat the same, and yet completely different. The diamond crystal is cubic in nature. Graphite crystals are hexagonal. The diamond is an insulator for heat and electricity, whereas graphite is a conductor. Understandably, the diamond has completely immobile electrons, its hardness factor. Graphite, on the other hand, is so mobile and soft it can be used as a speedometer cable lubricant.

Siegfried Marian, the engineer-scientist-agronomist who made the journey to the farm of Papa Sandor described in the Foreword of this book, published a soil magazine in 1949, 1950 and 1951, at which point publication was suspended because of Marian's death. Only a few pages of his work remain in the archives of the world. One paragraph is so rich with uncommon good sense, it rates in-depth examination.

There was probably a period during which life was capable of converting these two inert types of carbon, i.e., diamond and graphite, into structures and patterns which formed the basis of what we call life today [wrote Marian]. Today living structures are such in which both types of carbon exist together. Although both types have different functions according to their different characteristics, life is a function of the interaction of the two types of carbon which can best be described as a grid of conductive carbon [hexagonal] through which the system of life distributes its energy, insulated by the other type of carbon [tetragonal]. Life is thus comparable to a cable in which the live wire is protected by insulating matter.

When I first read these lines I must have let out a long, low whistle. I had studied the works of Julius Hensel, Sir Albert Howard and Lady Eve Balfour, and had ranged through most of the literature commonly called organic. A considerable body of work rejecting von Liebig's salt fertilizers had sprung up even before the end of World War II. Yet Siegfried Marian—in those fragmentary photocopied pages handed to me by a friend—rang a bell that still echoes in the far recesses of my mind.

Although I took to haunting libraries for more information on hexagonal and tetragonal carbon, that well seemed rather dry. But persistence pays off—in time. By the inch, even by the centimeter, the lessons emerged.

Tetragonal carbon in living things has a spin over the hexagonal carbon. This spin is announced as a chemical change takes place from solid to gaseous carbon, or vice versa. It is in this spin process that living things pick up energy—heat or motion—as the conversion is made from solid carbon to gaseous carbon, and plants convert gaseous carbon into solid carbon by drawing on the energy of the sun and cosmic radiation.

So here you have the spin of one type of carbon over a stable and conductive type of carbon conferring the status of life on otherwise dead charted elements. This life status can be damaged by a reduced spin, or some interference with the spin—or the spin may stop entirely, meaning death. Dead plus dead equals dead. It takes the spin of life to charge and vitalize life, and therein lies a modern dilemma.

The spin potential is reduced by an accumulation of waste matter from the process. Or, to put it another way, there gets to be an incomplete conversion between the two phases of tetragonal carbon. This conversion from solid carbon into gaseous carbon can become stuck in the horse latitudes of the intermediate state, meaning the conversion from solid carbon to gaseous carbon can bog down at the acid stage. The spin can be reduced almost entirely when the acids react with metals of small atomic volume (and dense electric fields), namely calcium, magnesium, iron, etc.

The position staked out by the carbon connection is simply that soil fertility is violated by a pileup of acidified or mineralized dead carbon. When the life giving factor or hexagonal carbon, namely biogenic carbon, disappears from soils, the end product has all the earmarks of inert coal.

Siegfried Marian hammered out the punch line to all this in December 1950. "*The future of soil fertility is a reversal of the process. Prevention of soil formation, keeping hexagonal carbon in circulation and extraction of hexagonal carbon from acidified and mineralized coal, peat, etc., must be the objects of the human race; if we are to prevent a complete breakdown of the human system, i.e. relying on our food from soil.*"

This secret eluded Marian, although he was well on track. It is a secret I have uncovered, and it has conferred on me a fuller understanding of soil health and human health.

Soil receives all of the toxins generated from wherever. Everything comes back to the earth and when it comes back to the earth; it often comes back in a toxic form. When the soil cannot adjust to this overburden, sick soil becomes the observed result. With a reduction of "live" carbon in the soil, and with the arrival of man-made chemicals as well as fertilizers, the health of the soil has become aggravated. Before the general introduction of this synthetic fix, the soil was in decline due to poor rotation management, and monoculture farming. Sick soil interferes with the welfare of bacteria, molds, fungi, viruses, all of which start to perform in abnormal ways. In the case of bacteria, they are often able to mutate to the condition, but molds are destroyed generally because they cannot get started in the business of increasing and multiplying. If the molds and fungi do not survive and proliferate, then the viruses become part of the structure, flourishing under abnormal conditions. There appears to be a direct relationship between certain energy fields and viruses. Viruses are to soil what the match is to a fire. They help correct an ecological imbalance. Viruses attack everything. They attack life forms, causing them to weaken. Then insect pests attack to continue the process. If the plants are strong enough to endure, life will continue. If not, death follows. Viruses definitely are a very important factor in the life cycle. The circulation of carbon in the life cycle between solid carbon compounds and a gaseous carbon compound requires the use of oxygen from the atmosphere. There can be no life without the reaction complex between carbon and oxygen, because the most useful bacteria are aerobic and need oxygen for their life and work process. As soil gets detoxified, bacteria start to function. Their very life processes cause the soil to crack. These are not the dry cracks of a drought. Generally these fissures exhibit long patterns, and are often several hundred feet longer than drought scars. This crack will go in a specific straight line and then section off into very erratic patterns with no distinct line. In health colonies, hairline fractures occur in the center of these now cracked areas. Bacteria need a tremendous inflow of fresh air to start the cycle of life and to get carbon going again, relying on oxygen from the air as food, leaving nitrogen to sink because it is heavier than air. In short, *bacteria feed on carbon and cause the soil to crack.* This makes possible the entry of air to soil. Bacteria can allow air to flow 100% in and 100% out. They have their hand on the switch, so to speak.

I have noticed that the good healthy colonies of bacteria will not let a waterway go outside of its banks. At our test farm we have water running down along the edge of the field year round. Generally you expect areas with that much water to be excessively wet, but this does not occur. The water is kept in its channel by the bacterial work force. At times it appears that a worm might cause a hole or a fracture in the wall, and bacteria might close it up in a very short period of time, usually in a few days, depending on the conditions. Bacteria, properly nourished with the access to the right carbon form, control the soil because they control the air coming in. Oxygen is required to convert carbon to a usable energy source, and carbon is needed to keep the oxygen coming.

Anaerobic bacteria, in return, require large amounts of nitrogen to convert matter to humus. Once the oxygen is used by the aerobic bacteria, nitrogen is left. This sinks to the anaerobic bacteria level and finds them. Microorganisms do not get oxygen out of carbon. All the oxygen that they need comes from the air. I don't think there is much oxygen in carbon, and it is not critical. Maybe after I have been following this line of study for thirty years I will think differently. It may be that there is an oxygen carbon factor that is critical, but at this point I have seen nothing to indicate as much.

If this were the case, then putting on large amounts of Nutri-Carb should stimulate the bacteria more aggressively. This is not the case. In fact, the opposite occurs. When we apply above optimum amounts, then we interfere with the ecological balance of the soil and bring on the very conditions that have developed with the use of the commercial fertilizers and fossil fuel pesticides. So it appears that the oxygen that is in the carbon structure is not a critical element. More critical is the ability of bacteria to generate air intake through cracks in the soil. It appears that Marian was looking for new answers to the old problem of balancing the soil. If bacteria, fungi and pests were the real problem, then life would have ended hundreds, maybe thousands of years ago. Quite the contrary, these phenomena are effects, not cause. The forest can feed on itself forever. Cultivated fields can only do this for a short period of time when the proper forms of carbon disappear as the desert complex takes over. Such soils cannot produce without steady infusions of new energy, usually salt fertilizers—the highly

MORE CAPITAL

How justified I was in turning my back to the propagators of "artificial" can now be seen in full evidence. Chemical salts as fertilizers were recommended over a hundred years ago. Over this period of "fertilizing" we would now expect such giant crops and such healthy conditions that we would have no cause for fears and worries. Year after year these fertilizers are poured over the land. Their "fertilizing" action, if viewed as a capital investment, should find us with extremely fertile fields and gardens, with exceedingly healthy crops. In other words we should expect a paradise.

There is no sign of any paradise. On the contrary, our fields and gardens are less fertile as time goes on. If these "fertilizers" were fertilizers, we would need less and less of them progressively. Instead, the sellers of these salts tell us we must put on more and more as time goes on. Troubles from insect pests and plant diseases have assumed terrifying proportions. To combat them, thousands of scientists are engaged on patchwork. They are not ordered to go to the roots of the troubles. They are ordered to find cures for this or the other individual trouble. So they cure a specified trouble with something which opens a new trouble. Patchwork, nothing more than patchwork, and we see the fabric of life from soil in rags, miserably supported by fading and ugly patchwork.—Siegfried Marian and K formulations. Then the "coal precursor" give way to the "desert precursors."

Perhaps it would be more correct to say that *artificial fertilizers aggravate conditions which are primarily due to the blocking of the soil body by coal precursors.* The coal precursors are exacerbated by negative energy bacteria. These special critters thrive on negative energy and caused the down-hill process to complete itself at a much greater rate. It has long been an academic conceit that we have to balance the soil and replace what we take out. In fact energy is never lost. It has merely changed into another form. It may be taken out through fission, which is the splitting of the energy unit. The opposite of fission is fusion, the joining of energy as in plant growth. A plant takes the carbon dioxide, the carbon monoxide and the trace elements and fuses them together into the tissue evident on, say, the corn stalk. That is fusion. When it is plowed down into the soil, it becomes part of the structure—the opening chapter, in fact, on how humus gets down there, and how coal is formed. Human diseases have their counterpart in the soil. Energy fields determine the character of a virus, that ever-present precursor of coal. A kind of raw humus, it kills life in soil, and in animals and man. Rheumatism can be said to be little more than coal deposited in the body.

All talk about energy, acids, alkalis, relate to human disease or faltering metabolic conditions. All acids are negative energy. There are also "like viruses" in the positive energy field, in alkali territory. Marian was probably well on track when he pointed out that all our ills and troubles are defined by different parts of our bodies stopping to have a carbon spin, and instead end up forming compost or humus or coal. We have to be careful when we talk about compost, humus and coal. They all suggest different processes. Compost, as we know it today, is something quite different from a related process in nature. One form of composting by nature results in peat. Bacteria can't use it or convert it to an energy source for plants. It's dead. That's a difficult concept for many eco-farmers to handle because a lot of composters use some of last year's compost to start this year's compost, exactly the way the old baker used last year's yeast to start this year's bread. There is always a hot ember in new ashes. Compost consisting of negative energy would create energy material only good for growing negative life. So compost may be dormant, but when it is ignited again, it goes to work. Energy fields must be available to ignite it. That's where carbon comes in. Carbon brings on a new source of organic matter. As a consequence the process can start all over again because of biogenic carbon.

It was the attempt to unlock hexagonal carbon that led to the development of actumus, a substance that approximated the content and quality of earthworm castings. Actumus was available in England and parts of Europe for a few years in the early 1950s, at which point it vanished along with the research, laboratory findings and records of Siegfried Marian. From the few scraps of evidence that survive, it has been possible to make a connection of sorts, and unravel the deep mysteries of hexagonal carbon. Marian had a thing he called biogenic extract of oak. It was for external use only. It had no taste or smell. It was a stable product and had all the biogenic properties of hexagonal carbon. It was recommended for use on the scalp to prevent loss of hair and as mouthwash, and so on. Marian was taking oak and distilling it until he had kind of a powder, perhaps on activated charcoal processed to a hydro form. He drove off the oxygen and other air elements—in other words, he ashed it, and then these ashes were used for poultices and rheumatoid complaints and inflammations and nose and ear applications, and so on. He really converted carbon into another energy unit. This new unit was added to a higher raw energy form with small units. This several decade old experience is of maximum interest because EPA recently found carcinogens attached to carbon dioxide, which previously was believed impossible. The lesson is clear. All toxins will attach themselves to carbon in any state.

When farmers come to my research acres, I say, *Welcome to the farm that reproduces your problems.* Marian knew that if he could take a problem and reproduce it, then he could come up with a solution. He took carbon, refined it almost to a hydro state, and used it as a poultice on human patients. This carbon form automatically absorbed the energy that was causing the problem. In other words, hair loss was nothing more than a consequence of degenerative cells no longer functioning because of toxins. Today hair analysis is a respected system for measuring debris parked in follicles by a system seeking to survive. In fact hair analysis is the one test athletes on anabolic steroids cannot defeat. The test reveals steroid use months, even years earlier. The Marian hair treatment was much like taking a tree in decline and giving it proper nutrients. The right carbon form may never bring the tree back to its original vigor, but it can prevent further deterioration, and perhaps initiate partial recovery. What is there about oak that makes it superior to anything else that you might reduce to hexagonal carbon? The answer may be citronin acid. All trees have high levels of negative energy. Therefore, opposite fields would be attracted in this arrangement. Marian lived in a time frame when investigators were coming to understand real negative energy in large numbers. Possibly he was able to get results that today would be unavailable because we have created an abnormal amount of negative energy, now in wide distribution. These are contaminants in the air, soil, rain, whatever. And they must rejoin through a type of fusion. Once they fuse back together and develop a new energy field, it then can attract the opposite side or join equals. We have driven a wedge between these two energy systems with all our pollutions. As a result we have an excess of negative energy, and this promotes poor health. If the soil is sick, then anything that human beings eat from the soil will be sick. Even degree and consequence vary. When human beings were able to adjust to this sick condition and have something that resembled health, the human body absorbed through the skin, through breathing and eating all the toxins.

Because the wrong energy fields are involved, the body is constantly in turmoil trying to adjust to that problem. This attempt has delivered a condition in which chromosomes and genes mutate, and the next offspring suggests better capacity to adjust to toxicity levels than any the parents had. Man is the only life form on the face of the earth that takes the weakest of two species and breeds them together with the objective of creating something stronger. This human penchant does not occur in unmanipulated nature at all. The human race murders healthy people in wars and encourages those with degenerative metabolic problems, with mental and physical problems to pretend strength from weakness. Society makes a great show of preserving such life. Nature never does this. There is a bottom line in any chapter of the carbon connection. *Life is a function of the cohesion energy and oxygen transmitting power of hexagonal carbon with the oxygen consuming power of paragonal carbon under formation of free energy, that is fractional energy.* We may look to hydrogen as being present in carbon as the mover of energy, with oxygen as the fuel. What does that mean? There are different energy fields within the carbon structure itself. This may seem technical and have no meaning to the average person. To put it all in perspective, we can state that without paragonal carbon, oxygen cannot be consumed and without oxygen, plants, animals and human beings cannot live. Biogenic carbon then, is the spark of life carbon, a leading actor in the cycle.

This is the area of absorption carbon as a purifier. Trigonal carbon contains a reserve of free energy potential and becomes the feeding ground for microorganisms. Bacteria causing alcohol fermentation may be considered the first step. Alcohol is poisonous to most bacteria. If alcohol-prone bacteria are killed, they are replaced by other bacteria which thrive on the alcohol, first making vinegar, the next energy field. In the meantime, hexagonal carbon still carries on the business of supplying the system with oxygen via aeration. This oxidizes the alcohol into an acid of paragonal carbon. Such acids poison further activity of the hexagonal carbon, and the system stops most form of life in its tracks. A microorganism complex without the appropriate level of carbon in the soil causes decay to go in the wrong direction. Instead of breaking down the lignens in the corn stalk, such decay actually accounts for fermentation that produces alcohol. Alcohol production in the decomposition process is for annihilating friendly microorganisms so necessary for soil maintenance and preservation of carbons. An acid condition in the soil is entirely a function of the carbon complex in the soil if carbon is saturated with negative energy (acid). Then a new carbon source is needed to correct the problem, or a different life structure able to grow under this condition. Parenthetically, it might be pointed out that—at one point—it would have cost billions and billions to get rid of PCBs, they were that indestructible. Now you don't hear about them anymore because scientists found a bacteria that mutated itself and it can actually feed on the stuff and destroy it. All that is left is a black gook. We take a black gook out of the ground called oil. PCBs are one component of that barrel of oil. Now PCBs can be placed back into the cycle by bacteria as a unit of energy. The role in the soil is really about the same as the role of human health, both in the transport of the right nutrients to the right location—whether it is a leaf or a limb—and its arrival there in a purified form. As the filtering material becomes depleted in the soil, it will automatically become depleted in human metabolism, and therein we find the precursor of bacterial and viral infections and degenerative metabolic disease. Yet there is more filtering material than agriculture can use, if we have the wit to crack its secrets. A perfectly valid carbon form is contained in coal—that is, in the right kind of coal.

Biogenic carbon is locked in this unit of carbon, but 99 1/2% has to be converted to humus to release the other 1/2%. That 1/2% is biogenic carbon, the spark of life. Biogenic carbon is indestructible. We can now speculate on how scientist Marian made his actumus. He put his materials in a barrel. He roasted it and allowed the gases to escape. If he hadn't allowed the gases to escape, he would have generated only an oily mess. Instead, he came up with a sort of crematorium powder, something on the order of the ashes left over in the little urn some people prefer instead of a coffin. And that's what he inoculated into the soil. Actumus started performing the function of cleansing, much as an activated carbon filter does in a still to take the bad taste out of water. Because the rainbelt soils of England were in fair shape, it worked. Today it might not work as well. Certain coals under the proprietary brand name of NutriCarb have repaired the shortfalls that mystified early investigators, and continue to baffle many still on the scene.

3

The Carbon-Oxygen Bond

Looking back, it now seems that von Liebig's salt fertilizer agriculture became sunset technology almost from the start. Sir Albert Howard, returned from India, touched off the debate with findings unheard of in academia. In one of his books, *Soil and Health*, he charged that two basic errors were sweeping the republics of learning, namely partial and imbalanced fertilization and toxic rescue chemistry. And with that the seed of a new farm technology was planted. It soon took on the name *organic*, but this nomenclature had little to do with synthetic construction of molecules around one or more atoms of carbon. In the late 1940s and early 1950s, the term of choice was humus, and the prime requirement for a healthy soil was perceived to be humus. Humus had been neglected for almost a century in Europe while killer agriculture embraced the several forms of N, P and K. The organic people blithely went about their way preaching the primacy of organic matter.

No one seemed to know how to define humus. It was created—if that word will do—by breakdown of organic materials, everything from crop stover to stable manure, tree bark, soggy leaves, heather and mosses included, substances that varied in nature, tone and content. Only a few observers paused long enough to reason that substances so different in origin and character couldn't possibly yield the same end product.

There were common denominators, of course. In this drift to raw humus, bulk was reduced, whatever the parent material. Moreover, from different types of organic matter, different grades of raw humus emerged. The addition of biodynamic mixtures, lime, charcoal, and the style of stacking and turning, all visited their effect on the humus end product. Those who studied humus discovered that the several processes of formation, reformation, dissolution, condensation, alteration in the molecular structures, all continued even after this humus was returned to the soil system. It was speculated that humus at worst retained some of the toxins in the parent material, and at best contained a balance of harmful and useful products.

The agency for this transformation was and continues to be trillions of microorganisms, with genus and species so varied they have not been counted to this day, and even the most astute investigators have only a hint of an idea as to which microorganism does what. The social structure of this population is governed by a number of factors, namely illumination, the chemical character of the soil and its oxygen content, moisture conditions, climate—and carbon.

Charcoal is made from wood. It always contains very small amounts of an organic compound closely resembling substances found in raw humus. This compound is humic acid. Humates, sometimes used in agriculture, are salts of humic acid. Humic acid itself is a structured carbon-oxygen compound. Its molecules are arranged as crystallites in the form of a woven mesh. This suggests the theory that the presence of this compound in trees is the result of those trees taking up soluble salts of that very compound during their life times as part of their metabolism from the soil, and of course the very same compound is found in soil humus.

When the tree—via the charcoal mechanism—is the source of this carbon-oxygen compound, the humus compound is always the same, regardless of the wood species used. When raw humus in the soil is handled by earthworms, the casts will have exactly the same amount of this carbon-oxygen compound. Thus the common denominator presents itself. Although raw humus—whatever the source—is never the same, the specific organic carbon-oxygen compound is always the same in all worm castings regardless of origin.

The conclusion, I think, is obvious. It is necessary to distinguish between raw humus which is unstable and the carbon-oxygen end product which is stable. The end product old Sandos gave Siegfried Marian with a command to figure it out must have been this carbon-oxygen compound. We can't be sure because the sample no longer exists, and Marian's records are not available. Marian called this end product *actumus*, a name that hasn't made it into the dictionaries and isn't being used in any forms of eco-farming that I know of. We can use it here to illustrate our points and to keep in focus this part of the carbon connection.

Plants require traces of this carbon-oxygen compound, herein called actumus. They need it for their metabolism, for which reason the Creator has gifted plants with an operating blueprint for taking up the substance from the soil solution so that it can be distributed evenly throughout the plant. This scenario suggests that when organic matter or raw humus is returned to the soil, it becomes the function of microorganisms and earthworms to free this carbon-oxygen compound from the bulk called raw humus for plant uptake. In other words, a shortage of the carbon-oxygen compound called actumus restricts proper metabolism in exactly the way a shortage of trace minerals inhibits growth, health and fruiting, and singly or in combined form deficiencies of either trace minerals or actumus become hand carved invitations for insect, bacterial and fungal attack.

In the Flint Hills of Kansas, there is a continuing argument over the matter of burning old pastures. It may be that when farmers burn most of the organic matter to ash, there is overcarbonation and a loss, but the burning contingent may not be

ACTUMUS

One of the outstanding factors noticed with the use of actumus is the enormous increase of root growth of plants in actumus treated soils. The cause for this phenomenon has been fully investigated, and it has been found that the root formation is due to the reactivity of actumus with the soil atmosphere and the soil solution.

Actumus is built on a crystal system of hexagonal carbon. On this system the electronic configuration is similar to aromatic organic compounds. On contact with a balance of moisture and oxygen from the air, actumus forms additional compounds of aromatic organic compounds. The number of such formed compounds is very large, but there is a predominance of one substance which is hexahydroxy-tripheno-quinone. This predominantly formed aromatic compound acts as very powerful root hormone, and the increase in root growth is thus fully explained.

This naturally forming root hormone on the crystals of actumus is a fairly stable organic compound. It can only be destroyed and inactivated by strong mineral acids, e.g., sulfuric acid, hydrochloric acid, etc., and by large quantities of free lime. This explains the fact why actumus is less efficient in the presence of artificial fertilizers containing mineral acids, and in the presence of excess lime. In natural soils or in soils which are cultivated without the use of artificial fertilizers and excess lime, the presence of actumus ensures a most remarkable root growth through the formation of "hexone" which is a short name for the above-mentioned root hormone.

Also forming on the crystal structure of actumus is 1-tyrosine. This organic compound is the life hormone of the earthworm, and the increase in the earthworm population in actumus treated soils is now fully explained. Both compounds, i.e., hexone and 1-tyrosine, are also exhibiting hormone action towards nitrate forming bacteria in the soil. These bacteria are strongly activated in the presence of these hormones. This explains the fact that actumus treated soils are never short of nitrate nitrogen.

Experiments with pure hexone and 1-tyrosine in the absence of actumus showed that they are effective without actumus, but their action becomes quickly exhausted, and it is necessary to give fresh additions almost daily. Actumus, however, ensures an automatic and continuous supply, because these hormones form as long as actumus is present as a result of reactivity between actumus and the soil environment. It is a continuous manufacture of these hormones as long as air and moisture are available.

The fact observed by thousands of users of actumus that transplanted plants, seedlings, etc., never wilt when treated with actumus is due to the almost instantaneous manufacture of the root hormone hexone which activates at once root growth of the transplanted seedling or plant.—Siegfried Marian so wrong after all. If it were possible to carbonize the old grass the way charcoal is made from wood, it would be possible to achieve a higher degree of actumus concentration than is generally possible with compost.

As noted earlier, carbon breaks down at the rate of 99.5% to humus, 1/2% to biogenic carbon. This, basically is the carbon to humus principle. Raw humus is really made up of organic matter which has been converted by bacteria to humus. Both anaerobic bacteria and aerobic bacteria figure in the process, the aerobes finally taking the humus and converting it into energy for the plant. Humus—on closer examination—is made up of the bodies of bacteria. It has a water retention ratio of 95% of its weight when in a healthy state. Harnessed with its water component, humus can be a fuel for the fire for life. Humus is another word for the bodies of bacteria. During their short life spans, bacteria have, in fact, consumed phosphorus, trace elements, all those nutrients that have become part of the body, and this explains how humus and bacteria become so closely related. Many organic folks believe that humus stabilizes the soil. This is an over-simplification. In fact bacteria stabilize the soil when they convert organic matter to raw humus. In the process they excrete a glue-like substance with a texture like urethane foam. It is this substance that causes the soil to separate slightly because it expands as it sticks to soil particles. It confers stability to the soil system. This is why a healthy soil simply will not erode. Air and water can penetrate such soil to supply the needs of bacteria. Soil has to have life to be healthy. And it has to have structure to entice air into its interior.

The carbon connection has an electrical connection. Phosphates and potash and other so-called basic nutrients are nothing more than electrically charged energy fields in rock formations. Each is a field unto itself. When bacteria take a bite of phosphorus, they are taking a bite of that energy field. It may be that a little explanation will permit improved understanding of that statement.

First, most of the energy used by plants comes straight from the air. Second, radiation has been used as a method of bombarding nutrient elements and exciting them to permit measuring. Radiation is often conceptualized as an electrical energy form. When used to bombard certain nutrients, this energy form excites the element. Computers can read the results and determine the level, say, phosphorus is on the energy field scale. Perhaps it would be correct to used the term *electric glue.* This glue gives formation and structure to soil particles, illustrating why bacteria are so important. The humus can only be considered a precursor to this life. And yet humus is essential for life's working arrangements. It is part of life. The other thing humus has revealed is that worms cannot survive without it. Humus is their raw material in the business of assembling those carbon-oxygen molecules. In the process they help control energy fields in the soil by consuming energy out of balance and refining it. Their digestion process does this, making assimilation by the environment more efficient. If there is a cold spot in the soil, worms work on such an area. If soil is moving out of balance, that is where worms tend to feed. The life of a soil system can be determined by the size of the worm holes. If there is inadequate bacterial activity, yet an ample supply of humus, there will be no worms. Without bacterial activity, humus will not fracture to provide the needed carbon-oxygen compound—the actumus, if you will—nothing to feed on. Worm holes will be very small. On the other hand, if there is a good humus structure, then the worms will eat a larger hole because they feed properly.

In one of the early issues of *Acres U.S.A.*, the editor rolled out a series of articles that describe how the carbon-oxygen compound mentioned here emerged from the mist of creation. Our civilization has inherited this valuable substance, but it is locked up in a carbon-rich substance called coal. Unfortunately this biological energy is much easier to release for heating homes and driving trains than it is for crop production. During the days and hours that saw the birth of the organic farming movement right after WWII, the vision of refertilizing our planet from stored reserves was alive and gaining ground. After all, the business of extracting actumus from younger sources had been proved out, not only by the old man Sandos, but by Marian and his followers. Guided by a dozen articles so out of the mainstream the professors refused to even read them, a handful of far-seeing people applied actumus distilled from forest and peat sources, and reported the results acre by acre, hectare by hectare.

This coal, this compost of once living matter, this highly condensed and concentrated humus, has been made inert and locked up by mineralization and removed from the life cycle. And yet its carbon-oxygen component remains packaged like some natural computer chip, ready to do the work of an electric condenser. The surface of its crystallites carries a negative electric charge. Since the soil solution carries positively charged ions, an electromotive force is created to transport water and soil solutions. It thus becomes the role of the carbon-oxygen compound to supply plants with moisture. Without this carbon connection, plants reach a wilting point, and then they die.

If the negative charge is kicked into neutral by acids, excesses of aluminum, calcium and heavy metals, the mechanism breaks down. But if, on the contrary, the carbon-oxygen bond is maintained so that it arrives in plant tissue, it regulates moisture, salt concentration, and at the same time props up the cellular structure to resist toxic substances, viruses, malignant bacteria, insects and fungus agents.

The chemist can readily identify what is being said here. This actumus, this carbon-oxygen compound, has two carbon atoms to share one oxygen atom. It is unsaturated. One carbon atom combined with two oxygen atoms, carbon dioxide, is a saturated compound. It is because of this lack of saturation that this special carbon-oxygen compound can attract gaseous oxygen and condense it on the surface—loose adsorption. Such a loosely held oxygen atom is easily given up to roots, cells, whatever. Without oxygen in the soil, root cells die and plants suffocate.

Coal is really a nutrient source combined with the carbon that we need. All we have to do is to release its vitality. It has taken me twenty-three years to find out what is wrong and what is right. There are certain types of coal which simply will not work with a soil system. These probably are the ones that many farmers have tried with poor results. The average farmer cannot take his farm out of production to do research. Too much activity in that direction will see him go broke. He hasn't the time, and yet many farmers seem to want to do that. They want to substitute their judgment for the directions that somebody else has worked out for years. In some cases they hear that so and so has used coal on the corn field. They find the nearest pile of coal, grind it up and put it on their acres and wonder why they've wrecked their soil. Science, art and insight have to background success in this area. Through the centuries, farmers have figured out how to burn away stuff not needed. For instance, if you take lime and apply it to the soil after it has been finely ground, it takes about two years for it to work. If you take hydrated lime, which has been processed and heated, it will work immediately. Farmers still in their early sixties will recall the old process of roasting lime in a field kiln. This was to break down the molecules to make humus. And then the farmer would put the fired lime on the fields. Visitors to China and other Third World countries report this system still in use.

All forms of carbon will eventually end up as humus. Healthy humus is the heaviest humus and therefore it will sink into the soil. The only way to raise it back up again is with a moldboard plow. Now if this is correct, coal could be the healthy humus that got away and is now turning up as a fossil fuel deposit. Many farmers have a problem with the moldboard plow. They see it creating a plowpan barrier, but the plowpan tells of a lack of healthy humus in the structure of the soil at that level. Sick soil is the cause, and it is useless to blame the implement.

It is this observation that brings us back to coal. Is coal formed by going through the digestive tract of the microorganisms or is it just simply a matter of organic materials, under severe physical compression? I speculate that it is both—digestive tract first, then compression next. Some people are going to argue vehemently with this premise. Based on what I have seen in nature, coal formations are depressed areas that occurred thousands of years ago. Using this theory of healthy humus sinking down, and taking worms into the equation over a long period of time—the process continuing and continuing—it caused a buildup of carbon. And then it became mineralized, condensed, concentrated, compressed—in short dead, and this vital carbon-oxygen material is held down there in a kind of a prison, it now being up to human intervention to bring it back into the life cycle. All we know for certain is that a vital part of our capital of coal came out of living circulation, and was deposited out of reach as far as direct biological use is concerned.

The inventory of carbon facts I have recited lead me to speculate that NutriCarb is the approximation of actumus in some esoteric ways. I say *esoteric* because what I have to offer is purely a theory. Suffice it to say that I am able to take what is known about the carbon-oxygen substance once known as actumus and draw a "T" account. On one side would be NutriCarb, on the other side actumus. I think actumus does only one, two and three and NutriCarb, being different, does one, two, three, four and five. Let's put something behind those numbers. NutriCarb makes humus 100 to 1, and actumus hasn't done this. Actumus stimulated the plant growth in all the early experiments. NutriCarb has done the same. Actumus caused more water retention in the soil because it stimulated bacterial activity. NutriCarb does the same thing. And there the similarity ends because actumus did not exhibit the absorption capacity of toxins that NutriCarb had been able to effect in the soil. Actumus in plant tissue was observed to form a protective filter mesh type of protection against the entry of viruses, bacteria, malignant diseases, and toxic substances. NutriCarb has been shown to deliver equal or better results. It appears that the small amount of actumus four decades ago would make it mathematically impossible to absorb the multitude of energy fields now in the soil. Quite frankly, many of these fields weren't even available then, or in the soil forty years ago. There are no electrical charges in NutriCarb to break up inert soils. This is done strictly by the bacteria. Otherwise you could take electricity—AC or DC—and charge the soil with enough voltage to cause this reaction to take place. There are a lot of experiments on record that in effect have made the attempt. Raw forms of energy can be destroyed that way, but the process will destroy our food for life. Now we have to analyze what were the results when this was done. Did it activate production? If it activated production, then there is a very simple explanation. They peeled off or removed from the surrounding structure by electrical energy the nutrients, thereby putting them into a moving form. This forced bacteria to do something with them. It's much like a welding rod. If you take a welding unit and use the proper heat, you can fasten two pieces of metal together. On the other hand, if you increase the amount of energy high enough, you can burn the two pieces of metal apart. The one thing that I have not seen is a follow-through on the nutritional structure of the plant after such an experiment. Most or all experimenting fail to test the plants to see if they are in balance nutritionally. Healthy soil will produce higher nutrition and be balanced at the same time. There are several trace elements which the breeds of today do not take up. The chemical systems of agronomy are just starting to look at the lack of trace elements.

Let's take hybrid corn. We know that it will not pick up selenium or cobalt because they represent two energy fields they are unable to handle. Now they are coming out with this new bovine growth hormone or stimulant. I just saw an article which stated that the nutrient value of the milk from these animals is not consistent. Scientists have already documented the fact that the nutritional value of milk fluctuates as a consequence of this hormone stimulant. By using it, man or animal could starve to death or be poisoned because enough balanced nutrients could not be assimilated.

For life the bottom line here is simple. There are coal deposits in various parts of the world that are available to regenerate soils. We should be making available this coal to the farmers because it is much more important as a source of the right kind of carbon then adding NPK or all other fertility elements. But it seems that there is no coal capable of doing everything. It takes a blend of coals. Some coals are so loaded with toxins, they cannot perform in the farm fertility arena.

I am fortunate, because less than one hundred miles east of where I live in Ohio are anthracite veins. As a consequence, I have been able to get considerable coal for testing. Just west of my research farm is the beginning of the lignite vein, and I have been able to get all the samples I want. I have analyzed them. The ones that work the most effectively are those found in bituminous deposits. Later on, as my research takes me further on down the road, it is conceivable that I may be able to unlock the real value in other coals. This will not be easy, and may take ten to fifteen years of research just to prove it will work. It would take several years of intense research to even figure out how to proceed.

4

Conversion of Energy

Robert Service, in his sad and hilarious poem—*The Ballad of Pious Pete*—tells the story of Sam Noot, "who started to shoot electricity into my wall," "my wall" belonging to the narrator, Pious Pete himself. Electricity was the ultimate of mysteries at the time, the energy of energies, capable of lighting up a Yukon cabin in even a small city. Some people considered it a dangerous plaything, the property of Satan, for which reason Pious Pete "settled Sam Noot," having laid his pistol on his scurvy, degenerate neck.

A point in fact, the Creator has been shooting electricity into our walls from the beginning of what we conceptualize as time. Our existence depends on light from the sun—and may seem disturbed by sunspots which play havoc with our communication system and which seem to author droughts, storms and floods on planet earth. The revenue of light we receive travels at the speed of light, 186,000 miles a second. Some of it is enroute only days, some of it centuries—light years, if you wish. At the Chicago World's Fair in 1933, the entire circuitry of the Exposition was closed by the light from the star Arcturus, light that had started its journey to planet earth forty years earlier. It has been said that we always live in the past, for the light energy we harvest was dispatched a long time ago.

Energy placed on deposit in the Archean age—1,200 million years before we built our towns and cities—is equally a record of the past. More important, it is the geological textbook lesson that enables us to understand what scientists mean when they tell us about the conservation of energy—this admonition that energy can neither be created nor destroyed.

The essence of life is energy. Life dies, but the energy it contains is preserved. It is contained in all matter, even in stone, even when it becomes static. In the case of coal, static energy can be unlocked with the agency of fire, chiefly for mechanical purposes. This represents a waste, ecologists point out, for we waste this valuable energy capital not so much to support new life, but to comfort "culture" for a scanty lifetime.

Yet energy never disappears. It comes to us as a gift of nature, with most of it invested in our soils. Coal was once living matter, interest—if you will—on free energy in the soil. We now know how this free energy was lost, measuring this "loss" in terms of our life spans and our curves of history. *Lost* is a relative term. Thus our preoccupation with finding the key to reactivating coal energy, not to drive steam engines, but to sustain farm crops and—finally—human life.

All energy falls into two categories, positive and negative. This is an oversimplification. In fact, there are many different energy fields. In toxic energy fields, all negative energy moves counterclockwise, and all positive energy moves clockwise. Only carbon in the soil can stop the energy fields from moving. It is the bacteria which allow the orderly movement of these fields in the soil. In my special test for toxicity, a counterclockwise motion becomes evident when an acid or negative field is encountered. Energy fields that are positive exhibit a clockwise spin. I have categorized these into systems number 1 through 5. These several systems will be found in both the positive and the negative fields. The less complex forms of life rely on a lower number of energy fields. In other words, the natural, open pollinated seed life form would rely on a category 1 energy field. A more sophisticated life system such as manipulated seeds would be a category 5 classification energy field. All energy fields must be related to chemicals. The chemicals that we used initially in agriculture belonged to category 1. Today, the newest chemicals coming on line are category 5 in character.

Category 1 chemicals are sprayed on top to kill a plant through translocation into the system. Category 2 chemicals reverse the procedure. They are introduced into the soil so they can destroy the target, the unwanted life form from the soil up. Category 3 chemicals are a combination of category 1 and 2. Category 3 preparations use chemicals with atomic structure capable of bypassing the normal plant checkpoints. They get into the system and destroy it, hence the term *systemic*. Category 3 chemicals of organic synthesis are growth stimulators. They force plants to grow excessively, in effect growing themselves to death. Category 4 operates in the reverse. It is composed of growth retarding farm chemicals. And category 5 is reserved for very sophisticated chemicals which are going to attach the toxins to carbon atoms for the purpose of confusing the plant's on-scene engineering system into taking in the carbon with its overload of toxins. When the plant does this, it dies. The big chemical companies do not settle for confusing the plants. They call these products *organic* because they are built around atoms of carbon. This meaning of the term *organic* synthesis confuses those who want to understand organic farming, which relies on the primary of organic matter and its ability to make the natural carbon cycle work. Hydrogen is the element in nature that moves energy. Oxygen is necessary for growth of the fuel carbon as a life producer. Too much hydrogen in a living system raises havoc.

Adjustment of energy exchanges in soil systems has been the objective of agronomy ever since Justus von Liebig parlayed his mistake into an international reputation. This adjustment was not understood clearly a century ago, and cannot be said to command in-depth understanding even today. The goal of tillage, for instance, is to increase or decrease a freedom of energy exchange for the crop acre. The application of humus has been

FREEDOM OF ENERGY

Energy never disappears on our planet. This is the divine part of our existence that we were given energy, and that the greatest proportion of it was invested in our soil, from which we were formed ourselves on basis of this investment, as a result of availability of freedom of such energy. But, freedom of energy can disappear and appear from various causes of nature. Take for instance coal. This was once living matter, created by freedom of energy in soil. Through external causes this freedom was lost—as we chemists know by high acidity of plant residues—and through other external circumstances—as we chemists know again by chemical substances formed in the course of fermentation—the residues could not decompose any further. Those masses of not fully decomposed organic matter were subsequently highly compressed and condensed during those hundreds of millions of years into what we call coal, which is nothing else but highly condensed humus in which energy is stored, but this energy has no freedom of transformation and exchange for living matter.

I venture to answer the question of what is life with a clear formula of freedom of energy in a pattern of matter which allows such freedom. Freedom of energy has quite naturally a significance which is related to the time factor. Energy can be measured by physical methods. This would mean quantitatively. There is then the question whether a certain amount of freely exchangeable energy is transformable in a time of x or 2x, or nx. This, it seems, depends on external factors like for instance radiation by sunlight, climatic conditions, etc.—Siegfried Marian pursued to give the soil a high energy capacity. Certain microorganisms require high oxygen availability and in turn improve the energy level of soil systems in which they live, and those microorganisms that live in the absence of oxygen usually decrease the soil energy levels.

When the soil becomes toxic, whether positive or negative, weeds grow because they respond to the wave lengths of energy programmed into their blueprints for life.

The chief determinant of soil energy levels in the chemical environment. That is why salt fertilizers are often more of a hazard than a help, and why the conventional idea that these salts supply plant food is nonsense. The salts can raise energy levels, and in some cases they become *enabling substances*—if such a term can be accepted—that permit plants to take up their requirements from the soil, albeit not from the salts.

Ammonium sulfate, for instance, can work well in high alkaline soils, and yet reduce to a vanishing point energy in acid soils, even pushing energy into a negative posture. Sodium nitrate will function in an exact opposite way, raising the energy level in low pH soils, lowering it in high pH soils. With the honorable exception of Dan Skow, the veterinarian, and Carey Reams, the health professional, few have paused to outline the characteristics of these salts. Most of the college texts on the subject are not worth the shelf space it takes to keep them.

Suffice it to say that highly energized life can be expected only from highly energized soils. Or—to put it another way—soils from which energy levels have disappeared are reduced to providing only marginal life and diseased life. The virus, namely the enemy protein that has been shortchanged in the contest of life, and therefore borrows—*sucks* might be a better word—energy from other energized proteins. In witchcraft, such a debilitating life form capable of bringing disease and death was called a *succubus*, but this language has never made the transition to biology. So settle for "sucks!"

The bottom line here is that healthy soils—soils capable of conferring health on farm plants—can be turned around by the salts and given host potential for insect, bacterial and fungal agents by altering energy fields.

Scientists who have synthesized new *organic* structures, ergo energy fields, have repaired some of their damage by getting crops—grown from hybrid seeds—to grow in these inferior energy fields. For my part, I have classified the seeds into five categories. Category 1 is a seed very close to nature's own open-pollinated variety. Category 2 is once removed. Category 3, twice removed; category 4, three times removed; category 5 must be reserved for the future. This seed will have to be marketed by a laboratory because this will be a very sophisticated seed life form, genetically engineered for the superfarm that will grow it. In this sorcery, nutrition is the forgotten kid on the block. Or, more correctly, nutrition is the hidden kid silently groomed to deliver a knockout punch. It may well be used to control populations and regulate "permission for life," to use a term often associated with the late C.J. Fenzau. Production of a seed that can tolerate starvation is already a reality to an extent, meaning the seed will be developed to grow only on what's available, the available nutrients to be governed by computer programs.

Acres U.S.A. has called such technology "killer science" and "frustration farming," among other things. On the other hand that journal has labeled science that works with nature "high science." The reader can be his own judge and—to quote *Acres U.S.A.*—needs no "reference to a higher approved authority." We know that oxygen absorbed by biogenic carbon has electrically charged vibrations in the ultra-violet wave length region, the region in which photosynthesis takes place. Moreover, the electronic behavior of carbon-oxygen compounds in hexagonal carbon crystals have been measured. Carbon-oxygen compounds are neutralized by acid fertilizers, sulfuric acid, some form of lime and many forms of organic matter in a state of partial decomposition.

These energy fields have many aliases. They include the full spectrum of classical energy, from X-ray to radiowaves, with lightwaves in the middle. But there are also energy fields beyond what is generally recognized. There are subtle, low level energy fields which are often seen as minor. In the field you can often walk into a ground vortex and feel the energy radiating out of the soil, and it is very cold. That may well be the cold vortex Siegfried Marian was talking about when he described cold spheres and the hot spheres in the soil. They, he suggested, were nothing more than energy escaping from the soil that can be actually felt. I live over what has been said is the richest known uranium deposit in the world. The government says that two rems of natural radiation are safe, and regulators have used this norm to establish the acceptable levels of radiation from nuclear plants. The two rems of radiation that I can measure have to pass through many, many feet of soil. I think this soil filter with its large number of energy fields is relatively safe, but will prove dangerous to life in the future. I would classify it as spare energy. The rems that emerge from a nuclear plant, they represent a more concentrated energy field. The soil a farmer farms has various types of energy fields. What the alert eco-farmer does when he uses the right blends of coal is deal with these energy fields.

The subtle energies—which may be bioelectrical energy forms—derived from solar sources and diffused in the atmosphere, in soil, in humus, and in all plant, animal and human organisms, have been described by many scientists. Pat Flanagan, the author of *Pyramid Power*, tells us that over one hundred terms have been coined to label this energy. Most of us know about eloptic energy, by Galen Hieronymus; odic energy by Baron von Reichenbach; X-force by the Britisher Eemon; and orgone energy by Wilhelm Reich.

They may all be the same thing. Wilhelm Reich had a term—bions—he described as "microscopic vesicles charged with orgone energy." He said, "They can be produced from organic and inorganic material by a process of disintegration and swelling up. They propagate like bacteria. They also develop spontaneously in the soil." This bion, this orgone unit is *characterized by a formula of tension and bioelectrical surface-change of the same basic pattern as that of organisms in a state of sexual excitation.*

The reader may speculate, as did the entire generation of carbon researchers of the late 1940s and early 1950s, whether we have in biogenic carbon the essential element for the transmission of what Reich called orgone. As present in soil, these orgone vesicles would be found in fertile union with hexagonal carbon. Reich was not allowed to stay on for an answer. The courts gave governmental agents permission to romp all over the U.S. Constitution and burn his books before they sent Reich to prison, where he died—"was murdered," his family believed.

Many have sought answers to soil problems with the radionics device. I worked with it for several years with the aid of an operator who had the equipment. Over the years I have made comparisons between radionic scanner results and those available from a conventional laboratory. According to conventional laboratory results, the soil was improving, and my scanner results said it was not improving. All the while the animals' health was improving. My operator was not able to provide numbers so evident in the health of livestock. We even ran a test some eighty miles distant to correct a problem in a poultry house. We were unable to get the desired results. This was a controlled test and we knew that the birds were sick. We had turned around 2,500 of the 10,000 in the house using Nutri-Carb. We had improved their quality and their health. We then turned to using carbon radionics, but we could not achieve the desired results. The operator who was doing the work was a very good operator, I think, because he was able to find an energy source in my fields, citronin acid, which is put out by the oak tree. No on else had been able to tell me what was wrong around those trees, but he did it with his scanner. It may be that I have seen four cases of poisoning in human beings resulting from the use of the radionics instrument.

Therefore I have decided to leave this line of investigation to others. You can alter electrical fields with radionics. The problem is, if you alter them in the wrong direction, you can cause the problem to get worse instead of better. For every action there is a reaction. As long as there is a reaction, most people see that as progress, even if it is deleterious. Too many negatives literally destroy everything, for which reason I do not like to mess around with these energy fields. Others may experience greater success.

The rest of the world is reaching more sophistication, as with this business of seeds with energy. When synthesis takes place, mutation of structure is involved. In my opinion, dwarfism in animal offspring can be associated with documented energy fields. If tampering with energy fields can dwarf animal offspring, then there is no question that it will dwarf the human being. We have to come back to the carbon connection. NutriCarb has both the positive and the negative fields in equal ratio. It is *the* stabilizing force. Carbon is the element of stabilization. It is also the element that is most unavailable or missing on farm acres across the country. My own tests reveal that most parts of the world are functioning with about 10,000 pounds of carbon per acre. These 10,000 pounds of carbon can only pick up ten pounds of an energy field per acre. Modern agriculture is terribly short of carbon. Some fifty years ago. U.S. acres had about 50,000 pounds of carbon per acre. And one hundred years ago, 100,000 pounds of carbon was the norm. This steady reduction of the carbon bank account is largely a consequence of so-called scientific instruction to use salt fertilizers. To function—as Ehrenfried Pfeiffer put it—the salt fertilizers had to be stabilized. This took energy and the process used up the humus in the soils. One pound of excess nitrogen consumes one hundred pounds of humus and one hundred pounds of humus is 50% carbon. Each pound of this excess nitrogen has used up fifty pounds of carbon. It simply converted it into an energy field that went into the air and became gas carbon dioxide. Some farmers use granulated sugar as a form of readily available carbon. Yet sugar is more a hydrocarbon which does not have the associated elements attached for stable conversion. Most of the sugar available to farmers as a highly refined carbohydrate has been grown with chemicals of organic synthesis. Not only it is not good for human beings, it is not all that great for the soil. Grown would present no problem except that sugar is a hydrocarbon once removed from liquid. If you take a spoon of sugar and put it in water, it quickly dissolves to liquid, and anything in that form has little attraction capacity. It would capture the worst toxins, that is those with a very high number of atoms per molecule. With a million atoms to a molecule as opposed to 100,000 atoms to a molecule, sugar carbon to attach to another substance is enhanced. Blended coal carbon has a capability of neutralizing or knocking out a broad spectrum of deleterious energy fields.

One illustration suggests itself. If you take a five pound magnet and bump it with a ten pound magnet, that ten pound magnet will pull almost all of the iron filings away from the five pound magnet. But that five pound magnet will not let all the filings go. It will still hang on to a percentage of them. Now view energy in the soil the same way. A very low grade magnet will retain only a small amount of whatever it needs as against the large one. Carbon is a master magnet, so to speak. It allows all of the excesses to be drawn into the carbon. Only the weakest of these excesses will remain with the structure and attached to it. When the Creator made this master magnet, He knew that it had to be removed after it had done its job. He gave that job to the bacteria. The bacteria eat the carbon and break that energy field. As soon as that energy field is broken, the bodies of the bacteria—which are now humus—become a new storage area for that energy. That's why we say that disease or sickness amounts to storage coal, an unusable type of coal. Only viruses flourish in such an energy field. This negative energy stimulates a virus to grow and become a blockage to the normal function of the organism. The system is intricate, yet simple. It becomes intricate when man interferes with it and attempts to spoon feed nature's requirements based on partial comprehension coupled to breath-taking arrogance. Energy fields and how they work can be understood only if storage of energy is understood.

5

The Storage of Energy

Some degree of review and reiteration is inevitable as we continue to make our carbon connection. That connection has already governed over 3,000,000 compounds in what science formally calls organic chemistry. Carbon also backgrounds another series of compounds called *inorganic*, although these are few in number. The difference was once etched in bronze, but this is now changing. For instance, organometallic compounds can be considered under either the organic or inorganic classification. Elemental carbon is restricted to two well-defined crystalline allotropic forms, namely diamonds and graphite. Less developed forms in terms of crystallinity are charcoal, coke and carbon black. Elemental carbon is inert, fairly insoluble in water, dilute acids and bases, and organic solvents. Graphite has a melting point of 3,500 C. for instance, and becomes liquid in that general degree vicinity.

It takes elevated temperatures for elemental carbon to combine with oxygen to form carbon monoxide or carbon dioxide. When subjected to hot oxidizing agents such as nitric acid and potassium nitrate, mellitic acid results. Of the halogens exhibited on the Mendeleyeff table of elements, only fluoride reacts with elemental carbon. At elevated temperatures, carbon combines with several metals to form carbides. Combined with oxygen, carbon forms three gaseous compounds. These are CO, or carbon monoxide; CO_2, or carbon dioxide; and C_3O_2, carbon suboxide. Back in the halogen department, carbon forms compounds known as the CX_4 group, X being fluorine, chlorine, bromine or iodine. At classical room temperature, carbon tetrafluoride is a gas, carbon tetrachloride is a liquid, the two other halogen-carbon compounds being solids. There are also carbon tetrahalides in this fantastic world of chemistry, the most important being chlorodifluoromethane, CCl_2F_2—freon.

The Creator has decreed that all plant and animal life be composed of complex organic compounds, all containing carbon combined with hydrogen, oxygen, nitrogen, and the other elements associated with life—plant, animal and human. The remains of well-dated plant and animal life turn up routinely as petroleum, asphalt and coal. Natural gas deposits contain compounds composed of carbon and hydrogen.

These distinctions made, we can now pause to consider free carbon, which is available in many large deposits of coal. This is an amorphous form of the element. It contains complex carbon-hydrogen compounds, and energy generally unlocked only by the agency of heat. And yet there are coal forms that respond to biological activity and, perhaps, to the fantastic world of biological transmutation outline by the late Louis Kervran.

Industry has not been tardy in harnessing these several carbon sources to its technical pursuits. The diamond is both the god of jewelry and the epitome of cutting capability. Carbon black is an integral part of tire rubber and printer's ink. Graphite is used for high-temperature crucibles, arc-lights, dry-cell electrodes, office pencils and lubricants. Charcoal is an absorbent for gases and impurities, and as a decoloring agent.

Even the gaseous compounds that rely on a carbon connection have been industrialized. Carbon dioxide is used to put the bubbly in beverages, to extinguish fires—and in the solid state, to put out fires. Carbon monoxide is used as a reducing agent for many metallurgical processes. Carbon tetrachloride is used in fire extinguishers, but it—and carbon disulfide—are important industrial solvents. Calcium carbide is used to prepare acetylene for welding and cutting, and other carbides are used as refractories and metal cutters.

Dry ice is nothing more or less than solid carbon dioxide. It is obtained commercially as a byproduct of fermentation, combustion of carbonaceous fuels and certain forms of chemical processing. It sublimes directly from a solid to a gas, is nontoxic and noncorrosive. It is used as a shipping refrigerant because as it "melts," it leaves no residue. John Hamaker has associated carbon dioxide buildup with a possible impending ice age.

It is the quasi-biological transmutation—if we can coin a concept within a phrase—that interests us here, because carbon from coal—worked on by molds and fungi and bacteria—make real the proposition that stored energy can be freed, and free energy that is used up does not disappear, but becomes stored energy.

The storehouses for carbon are many. And the nutritional flow is closely guarded by nature, the eye of the needle being exacting in its demands for the right of passage. It matters not how high the nutritional structure may be served with a mineral fix. This fix cannot support plant growth without the catalytic action of carbon and its compounds. Carbon compounds supply the action by which the mineral component is harnessed and made active. It is carbon that gives permission to life. It is carbon that activates the nutritional flow of life. In fact there can be no life without both a minimum of zeoliths and their nutritional surface with a maximum of nutritional surface. Suffice it to say that carbon compounds with the right framework are constructed on hexagonal crystals. Inactive carbon compounds are build on the cubic diamond system. The celluloses, starches, fats and minor elements in residues of life lack the framework and nutritional surface so necessary to govern the flow of nutrients. But carbon compounds with framework structures, hexagonal carbon, if you will, have the cavities, tunnels and passages in their crystals to activate this flow.

RESIDUES OF THEIR PREDECESSORS

The British Isles were once completely under forest. It was not so long ago when the land started to clothe itself with trees. It is less than 10,000 years since the last ice sheet covering the greatest parts of the country receded and gave way to warmer conditions. On such barren land as was left by the ice, life started with small organisms for a meager existence. On their residues plants developed until trees came up to live on the residues of their predecessors. We call this natural succession. At that time there existed a land-bridge between these isles and the Continent; small animals, birds, wind, carried tree seeds from the Continent of Europe into the wastes, and soon luxurious tree growth developed on the warmed land. However small life is, it will always warm the land. Even microscopic fungi contribute to an increase of the soil temperature.—Siegfried Marian

Fertilizers fail, and nutrients will not travel without a water supply in the zeolithic-hexagonal carbon structure. But when water flows there is also a flow of nutritional elements. That is why irrigation of soil without this totally vital supply system is much like pouring water into the sea. And adding soluble salt fertilizers to soil depleted of their zeolithic-hexagonal carbon is like dumping sand on the beaches of Iwo Jima.

The charge is often made that we are using our energy at such a great rate that it is in danger of being exhausted. But in each case the experts have failed to take into consideration the fact that all energy is not lost. In fact, none of it is lost. It is merely converted into another form. When energy fields become dangerously high, then the membrane—the ozone layer we call it—opens up like an escape valve, and then closes up to stabilize the energy so that there is no huge explosion. I frankly do not think that ozone holes have much to do with fluorocarbons. If you blow up a balloon and find its weakest point, it bursts and all of the air goes out. I think we are releasing too much negative energy. All energy is stored in three areas, rock, liquid and air. Water is simply a neutral energy field. Phosphorus or potash that is buried start to convert to a hydrostate, or a liquid form. Sugar which goes from a liquid to crystal put into coffee becomes liquid again. So we have to used the word *hydro*. We can't use the word *water* because we are talking about a specific energy field. In a rock form, energy does not move at all. It stops. Over the years, *Acres U.S.A.* has published data on what each acre of land contains in those top six inches of soil. The values are available in a rock formation. Nitrogen represents about 2,700 pounds to the acre in a metallic form, which is not available to anything. Yet bacteria can assault that metallic form of nitrogen and convert it. Rock, then, is one area of storage. In an area where the glaciers concluded their march, stones exhibit distinct lines where they were chipped away from existing rock strata. The character of such rocks is quite different from those found further north, where stones were rolled and worn into a semi- rounded condition. Here we find rocks that have distinct edges on them.

COMPOSITION OF SOILS

For 1 Plow Acre 6-2/3 Inches in Depth
Approximately 2,000,000 lbs., or 1,000 tons

Elements	Sandy Loam lbs. per acre	Silt Loam lbs.per acre	Clay Loam lbs. per acre
Organic Matter	20,000 =	54,000 =	96,000 =
Lbs. of Nitrogen	1,340	3,618	6,432
Live Portion (Earthworms, Bacteria)	1,000	3,600	4,000
Silicon Dioxide	1,905,000	1,570,000	1,440,000
Aluminum Oxide	22,600	1909,000	240,000
Iron Oxide	17,000	60,000	80,000
Calcium Oxide	5,400	6,800	26,000
Magnesium Oxide	4,000	10,400	17,000
Potash	2,600	35,000	40,000
Phosphate	400	5,200	10,000
Sodium Oxide	4,600	26,000	24,000
Titanium Oxide	13,600	18,000	14,400
Sulfur Trioxide	600	8,500	6,000
Manganese	2,500	2,000	2,000
Zinc	100	220	320
Copper	120	60	60
Molybdenum	40	40	40
Boron	90	130	130
Cobalt	50	50	50
Clorine	50	200	200

Compiled by J.L. Halbeisen and W.R. Franklin.

From rock to availability covers a multitude of sin. In the laboratory the scientist takes a sample and subjects it to extracting solutions as decreed by official manuals. Unfortunately, plants cannot go out and buy those solutions for the purpose of extracting nutrients. A living, dynamic complex known as the living soil must do this. Here is what that life system has to work with.

We had dug 10,000 feet of ditch for a neighbor one time, and he commented on how he had hauled off tons and tons of stones every year. Yet when we trenched, we found few of the stones present in this 10,000 foot ditch. We had gone down as much as eight feet, and yet we could not find the stones he was picking up on top of the ground year after year. His belief was that these stones were coming out of the ground. I think that all of the potash, all of the nitrogen, all of the phosphorus, and the other things we have been placing in this ground together with all of the chemicals applied, have all become part of the new rocks that my neighbor has been hauling off.

We know that the stones contain a lot of minerals. If we read correctly what John Hamaker says about remineralizing the soil with crushed gravel, it refurbishes the theory I have illustrated. Over the last three years I have noticed a significant deterioration of stones. When I talk about stones or rocks, I am not talking about something two feet in diameter. I am talking about something one inch in diameter, a half inch in diameter, and up to four inches in diameter. I have photographs that I took circa 1989 showing how rocks actually disintegrate and they spread themselves over quite a wide area. Something occurs to cause these rocks to spread out. Visualize a four inch rock spreading out to eighteen inches or a half inch rock spreading out three or four inches. That means some significant power was present to throw pieces of stone that far. There is a natural breakdown of minerals from the rock formation by nature.

The next area of storage is the hydro form. We are led to believe that we must promote soluble fertilizers. Yet unless it was designed to grow on liquid soluble material, a plant will not grow on it. A plant that is designed to grow on liquid soluble nutrients will accept foliar feeding. On the other hand, if a plant is not of that design, you can put all of the foliar materials on it that you want, and nothing will happen unless a stimulator is used. I have documented this at my test farm. A number of farmers who had been using foliar feeding programs found out that I was correct. There is no benefit to a healthy plant. However a subclinically ill plant will benefit. Hydro storage is only one area of storage. Water is the only element on the face of the earth that can accept an acid or lime because it is a pure neutral dilutant.

As a soil becomes saturated with chemicals of organic synthesis or salt fertilizers, these materials migrate into the water, but this is only a storage area and not a process. The hydro process goes from one phase to another phase. But water does not go from a rock to a gas. Water either stays liquid or is taken into the air as a gas. Ice is not a rock. It is merely a formation that occurs because of the temperature. Stones, in turn, are not generally created by temperature.

The third area of storage is gas. Two numbers are pertinent here—2 and 98. These are percentages. Only 2% of what it takes to grow a plant comes out of the soil. Fully 98% comes out of the air. This means that plants are grown largely on gases. Aerobic bacteria in the soil convert humus to a gas. That, in fact, is what the plant grows on. Once that spark has occurred in the soil, then the plant can recognize it in the air and extrapolate it from the air for the process of fusion within the plant. The prime gas is carbon dioxide with a small amount of carbon monoxide attached to it. If you haven't the carbon, then you can have the wrong kind of gas. It might be butane or another gas not associated with the carbon latticework. The air that passes over the United States is very similar to the air that passes over the Sahara Desert, yet there is no plant life to speak of in the Sahara Desert because there is no carbon. An oasis has water and it also has continuing plant life. The desert caravans that stopped at the oasis allowed the camels to drink. Those animals defecated and urinated there. This was part of the process of the beginning. When I say "the process of the beginning," I mean the beginning of the maintenance of an oasis. If an oasis was allowed to mineralize entirely, if you took all organic matter out of it and left nothing, it would become much like the surrounding desert, nothing! It has to have water and it has to have carbon. The cycle must be complete. Wherever the cycle is not complete, life ceases to exist. Ironically, gas is the greatest area of carbon storage because it takes up the greatest amount of volume. The carbon connection with plant growth via air was well illustrated in *An Acres U.S.A. Primer*, which is quoted here with permission of the publisher.

Jan Baptista van Helmont, a 17th century Flemish physician, started getting a handle on exactly what happens when he performed his now famous tree experiment. He simply wanted to know how soil matter was being displaced when plant life grew. No one could measure such a proposition in a field, or in a forest. So van Helmont planted a willow tree in a large earthen tub. The little sprig weighed in at 5 pounds. Soil used in the experiment scaled in at an even 200 pounds. The tub was then covered so that only a small hole for the tree trunk and one for watering remained. Five years later the tree was not only larger, it now weighed 164 pounds. Obviously, reasoned van Helmont, if the willow tree picked up the difference between 5 pounds and 164 pounds, then the soil remaining in the tub should weigh only 41 pounds, potting material having been reduced to oven dry soil for the post-growth weigh-in. The results proved van Helmont hopelessly wrong. After contributing to the tree's growth for five years, the 200 pounds of soil had lost only 2 ounces. Van Helmont pondered the problem in deep consternation. Could it be that all this growth came from the water he had given the tubbed tree all these years? Surely this was the answer.

No, said a scientist by the name of Nehemiah Grew. He had used the new tool called the microscope to study the tiny pores of a plant leaf. Here were openings not unlike those observed in the skin of an animal or a man, obviously for the elimination of perspiration. By this time it had become uncomfortable to use the term *obviously* too much. Grew took a second look and this suggested another possibility. Perhaps the pores were for the admission of air. Van Helmont had completely overlooked air, and the willow tree had been in contact with air as well as water and soil. Even so, he didn't pause to consider that air contained dust particles—debris from volcanoes, pollution from forest fires and industry, soot from town chimneys.

It was one thing to have a vision of how nature works, another to test it out. How, indeed, could one determine whether air figured in the growth of a plant? An English preacher and scientist named Stephan Hales reasoned it out and constructed his famous peppermint plant experiment.

The proposition was hellishly simple. A peppermint plant was potted in a small cistern. Next, a glass was clamped down over the plant to shut off outside air. The water level in the glass that surrounded the plant was regulated simply by siphoning away water in the cistern. Now if the plant made use of the air, why the water in the container would either rise or fall. Hales knew about atmospheric pressure, but his control would handle that variable. The control was a similar setup, except that it contained no peppermint plant. Atmospheric pressure would treat each vessel alike. If a difference developed, this would be sound scientific evidence that the plant had interacted with the air in the vessel.

Hales tabulated his results and watched. One day the plant took on a noticeable wilt. Hales replaced this plant with another, taking care to preserve the old air in the vessel. It took less than five days for the new plant to fade away. Yet a plant introduced into the control vessel lived approximately as long as the first one. Apparently plants were taking something out of the air. Was this something food?

Of course! Once the food in the first vessel had been exhausted, a second peppermint plant couldn't live. Yet a peppermint plant placed in the "stale" air of the second vessel lived as long as the first plant. Being an astute observer, Hales reasoned that his mint plants were changing the air. How? He died without finding out.

The idea of using inverted glass vessels to follow where logic led didn't die with Hales, however. Joseph Priestley, the British chemist who was later to isolate and describe the properties of elemental oxygen, was fascinated by those trapped bodies of air. At one point in his experiments, he clamped such a glass vessel over a burning candle, and of course the flame went out in a short time. He did the same with a mouse and found it could live only a short time.

A mouse and a candle! What could they possibly have to do with agriculture? Priestley continued his experiments and finally announced a serendipitous discovery. "I flatter myself that I have accidentally hit upon a method of restoring air which has been injured by the burning of candles, and that I have discovered at least one of the restoratives which nature employs for this purpose. It is vegetation."

Indeed, even a candle consumes a gallon of oxygen a minute. People and animals used air with reckless abandon even before the advent of the automobile and the industrial age. Something was renewing the air, and Priestley stumbled on some parts of the answer. The recorded date for this discovery was August 17, 1771. First Priestley consumed the air in a chamber by burning a candle. Then he inserted a peppermint plant. It lived. Moreover, after only ten days a candle would burn in that same air again. The same principle held for a mouse. Air that had been exhausted until a mouse died could be renewed by the mint plant so that ten days later another mouse could live in it.

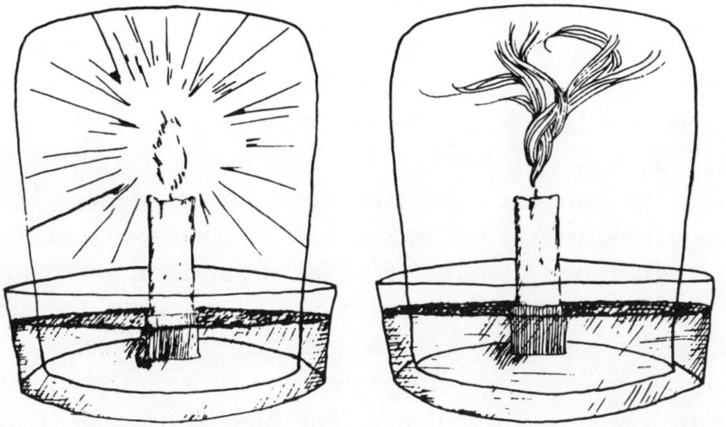

Priestley consumed the air in a chamber by burning a candle. After a candle had exhausted the air, a mouse could could live in it.

After a peppermint plant had renewed the air, a mouse could live in it until the air was again exhausted.

This seems so fundamental now that we wonder aloud how such complete simplicity could stand as a turning point in agricultural science. Certainly Priestley believed plant life was working this strange miracle, but he didn't understand why. At the time he worked his candle/mouse experiments, he hadn't isolated oxygen yet. As a matter of fact, Mendeleyeff's periodic table of the elements wouldn't be published for another hundred years. Because of this, perhaps, a final answer to the riddle being served up by Hales and Priestley had to wait well in excess of another hundred years for solution.

All of the materials used over these many years to replace or rebuild soils have come from rock. Much of our agricultural phosphate comes from Florida. It is simply rock. Agricultural nitrogen is taken from the air with a great expenditure of natural gas. Calcium comes from lime rock. Too frequently application of nutrient materials causes an imbalance. There are eighteen tons of lime per acre at 6 2/3 inches, calcium and magnesium combined. There are 44,000 pounds of potassium, yet only 2,500 pounds of phosphorus. If you want to balance a soil, the numbers have to be adjusted in terms of base exchange capacity. Potassium to phosphate has its ratio, and violation of this ratio creates an imbalance. The highest number of any element is carbon. The second highest element, believe it or not, is iron. When I test NutriCarb, I find that the second highest component is hydrogen, the first being carbon. In putting on a fertilizer, we should put on 60% carbon, 15% iron and then each one of the elements below it in pecking order. All energy is stored, one way or another. The fear that we will be running out of energy is born of primitive knowledge. We have only to figure out how to take gas from the air for conversion and use as energy. The agency for the most efficient conversion of energy in the air is the plant. That will not change. It would be possible to run van Helmont's test in the 1990s and the findings would be exactly the same. There has been no change in how nature works. If instruction out of the land grant colleges was correct, then farmers who have not larded fertilizers on their acres for thirty years, or three hundred years, would all be standing in great big holes, but this is not the case. *Farmers of Forty Centuries*, the story of Chinese agriculture, is a record that can't be refuted. But we can't get our university people to read it. Extension, in turn, fails, because its direct line goes to the same source of misinformation, the university. Our attention must be directed to energy stored in the three places defined here. This means we must take an added look at the billions of unpaid workers in the soil that perform for us that activity of releasing that energy.

6

Bacteria—Producing and Protecting Them

Dan Skow, the veterinarian and soil consultant from Fairmont, Minnesota, usually tells his farmer clients that carbon is vital, "and the depth of the carbon in the soil is the depth of the magnetic field." When real estate is changing hands on the wings of the winds, it is because there is no magnetism in the soil, and no live carbon. Inert soil carbon is converted to useful carbon slowly by the action of microorganisms. It is the function of microorganisms to consume the part of carbon that has no energy potential, and leave the other part to perform its vital functions.

This microbial assault on carbon represents a digestion process. This process takes soil carbon and breaks it into carbon dioxide and water, leaving other essential carbon compounds in the soil to pump water. This digestion process has been compared to the digestion of food by animals. They take in raw organic matter, extract energy from cellulose, convert it into gas and moisture, and leave behind some of that organic matter in the form of excretia. In animal manure farmers find a certain amount of carbon in its liquid phase, and this has the fertilizer value of soluble humus. Digested organic matter is ideal, but it is not always easy to know which of these materials are a good energy source.

Soil microorganisms like lignin. They get it by consuming the cellulose part of organic matter. Lignin is build on a three-dimensional system of carbon. Cellulose is built on planar or two dimensional lines. The pattern of lignin compounds is similar to the arrangement of carbon in charcoals. After nearly a lifetime of considering the matter, Siegfried Marian correctly concluded that fully digested soil carbon consists of lignin only and this lignin has further been dissolved into carbon compounds which are capable of extending into a watery solution similar to the liquid portion of animal manures. From such carbon only can we expect energy action for pumping water. This energy develops from reactions taking place between such liquid carbon compounds and soil elements.

We say all this not as a codicil, but as a preface to fuller comprehension of the role microorganisms play. This interplay between solid, liquid and gaseous carbon phases is a condition for permission to life, not so much chemically, but energetically, even though such energy is the consequence of chemical reactions.

It should present no great puzzlement, then, to announce that bacteria produce the energy necessary for a plant to grow first, and then protect life.

The microorganisms have handles. They are aerobic and anaerobic. They can be identified by genus and species in a loose sort of way. They can mutate to survive the rigors of temperature, high and low, and they can strike for higher wages in the form of suitable materials on which to work. The aerobes can be found in the top few inches of soil in the main, but they can also endure much lower than that if there is something to sustain life, and if there is enough oxygen. The greatest need is oxygen, a 21% component of air above ground.

The anaerobes convert organic matter into humus. I have seen anaerobic activity as deep as eighteen inches below the surface of the soil, albeit only when the soil was healthy.

Almost any farmer will remember plowing up corn stover that has been in the soil a year or more, apparently untouched by bacteria. A lack of nitrogen is generally identified as the chief cause, and this can't be denied. However, reason must intervene. Air is approximately 78% nitrogen, 21% oxygen. Nitrogen being the heavier of the two. When oxygen is used up by aerobic bacteria, nitrogen then sinks down to where the anaerobes live and give a super charge to their food supply. Organic matter converted to humus by these billions of workers is always balanced—well, not always. These microorganisms can become mutated to toxic conditions. Overwhelmed might be a better term, in which case the end product would be less than balanced. The fine line between anaerobic and aerobic bacteria is totally under their control. The anatomy of healthy soil maintenance is governed by the microorganisms and how they live and die.

Once aerobic microorganisms can no longer sustain life because soil tilth has been annihilated, anaerobes will thrive in their airless environment for some time before they too succumb.

Let's consider the protection of life. As high levels of chemical toxins—fertilizers, pesticides or so forth—are introduced into the soil, it becomes the function of bacteria to shelter the environment from these foreign invaders. It is reasonable to assume that microorganisms have to protect the environment first. If the environment is not protected, then there will be no reason to produce, for no natural life can exist. When microorganisms are in a "protection mode," they produce energy—that is chaotic energy, energy out of balance. When nature has ordered up this "protection mode," and imbalances the status of produced energy, hybrid seeds, weeds and grasses proliferate, products of a sick soil, all of them. In this protection mode, anaerobic microorganisms process toxic materials locked into organic matter. This processed product is then either attached to rock surfaces, or takes on a hydro form, or becomes gas which moves off into the air. That is how energy is put into any one of three areas for future use.

STRUCTURE OF LIFE

I know that you sometimes think my views are somehow radical and sharp, and I fully appreciate that you must find it difficult to see the necessity for keeping up this aggressive attitude. You can believe me that I find it very difficult myself because I am not an aggressive character. But, if we want to build up a new structure of life and knowledge, we must first profoundly clear away the rubble from our ground. I have used a kind of mental pneumatic drill on myself for many years. This drill has bored and destroyed all the waste of wrong conceptions. I could not otherwise have found a clear ground on which to re-build. Like the dentist who hates to use the drill on your tooth, so must I clear the waste of wrong conceptions from your minds before you will be able to take full advantage of the new possibilities. This picture of my work came very clearly to me the other day when I stopped for a few moments to watch pneumatic drills at work on residues of buildings which were destroyed during the war in Exeter's High Street. These entangled conglomerates of concrete, steel and brick require some hard boring, but it must be done in preparation for new buildings to be erected.—Siegfried Marian

If toxicity is so great that microorganisms cannot go into a protection mode, they simply die off, and their function ceases to operate. Now we have hard, dead soil. Many farmers can relate to that situation because there is a place or two on every farm where soil is dead, dead, dead. A plow cannot be pulled through such soil because it has the consistency of a solid cement block, the benchmark of a totally toxic soil. A totally toxic soil will still produce lower phylum weeds and grasses, nature's bid to repair the abuses of mismanagement.

It may be a defensible generalization to say that microorganisms cannot produce and protect at the same time. Anaerobes are producers when they take organic matter and convert it to humus. In doing this they take trace nutrients from rock formations or from hydro source materials which they need to balance the humus which they are going to produce.

Added to the life complex in the soil are molds and fungi, which help with the process of destruction. What part these totally different life forms play is not entirely understood. It appears, however, that the molds and fungi have the capability of growing on toxic energy. If toxicity is too great, even these are unable to thrive. Since anaerobes function in the absence of oxygen, and a chisel plow pulled through the anaerobic zone will raise the soil and create pockets of oxygenated air, such a procedure could prove detrimental to anaerobic bacteria—and to formation of humus. Many farmers will argue by saying, *We get results!* That may be correct, because as bacteria are killed, their bodies become a source of energy for a short period of time. But this is a short lived benefit. Once all the bacteria have been consumed, there is nothing left. You have the equivalent of the non-living desert.

The aerobes do not function in a protecting mode. They are merely producers. If the environment is too toxic to permit aerobes to function, they die out. Such a situation becomes the frosting on the cake called desert. Aerobic bacteria convert humus to an energy source based on the source status produced by anaerobes. That is how the aerobes produce the energy called for by the plant, be it a weed, a corn stalk, wheat, whatever. This means the aerobes produce that vital 2% of growth factors necessary for 98% extraction of growth factors from the air.

Carbon is to the plant life what heat is to eggs for incubation and hatching. Here I simply have to repeat that carbon is 50% of the humus structure. Carbon is broken down from carbon to humus. Then the carbon, which is now in a semi-hydro form in anaerobic bodies, is now attacked by aerobic bacteria. They convert this semi-hydro form to carbon dioxide, thus carbon dioxide becomes the biogenic carbon, the spark of life we have talked about. As carbon dioxide is produced, excesses escape into the air. These several points considered, we can now put into clear focus what Siegfried Marian called *actumus*, a product he created via a heat process that broke the carbon structure of certain forest materials into a humus type material. This process made it more available for the aerobic bacteria function. It was always put *on the* soil, and it never seemed to work when introduced *into the* soil at any great depth. It was worked into the top three inches. When heated, this carbon form broke down from solid to hydro, and became very much like hydrated lime. In lieu of anaerobic bacterial activity, and with the availability of a good source of humus, by manipulating organic matter he broke it down through a heating process to a point where aerobic bacteria could use it. Sustaining the aerobes became the focal point of Marian's exercise. We are led to believe that when the air temperature drops below 28 F., bacterial activity ceases. If that were true, one should be able to separate bacteria into anaerobes and aerobes. I have never known of anyone doing this, and have never tried it myself. I merely suggest that once the temperature changes, aerobic bacteria have the ability to convert over to anaerobes. In the spring, when it warms up, anaerobes have the ability to revert back to being aerobes and during this some destruction of organic matter can be observed, but because of the short time frame involved, it is not complete enough to sustain life. If it were, life would continue without change. But we know that fertilizers are now needed, or no production will occur. I believe that aerobes and anaerobes are, in reality, the same stripe. They have mutated to supply two separate needs in the environment. I have found bacterial activity on top of the soil at a zero degree temperature. If the two were truly separate, this would suggest movement—but the phenomenon appears almost instantaneously. By instantaneously, I mean over a 14 day period. I have never heard anyone suggest that there was a bus that hauled these microorganisms six inches down to the top of the soil. This occurred with one-half to one inch of ice following an ice storm. I believe the ice caused oxygen depletion. When this happened, it manufactured the right climate for anaerobic bacteria to flourish. Once the ice melted, activity stopped.

There is nothing in the literature to suggest migration. And it does not appear that anaerobes have the ability to do it. *Acres U.S.A.* has often reported on bacterial colonies. These colonies have an infinite ability to convert to anything necessary to correct the situation, which may be the reason for being of colonies. If the microorganisms did not live in colonies, there would be no continuity of movement in the soil. A group would move off and establish itself, but this does not appear to be the situation. When we first started using NutriCarb at my test farm, we had areas that were green and areas that were not so green. Soon we noticed that these areas merged. On-scene observation suggested that the green area represented a lot of micro-activity and these microorganisms were producing a lot of energy, consequently green plants and good growth. As microorganisms expand in numbers, they merge. Mutation of bacteria has become the darling of genetic engineers, but the fact is nature has mutated bacteria to consume PCBs. Bacteria can consume the bed of a truck and take atomic waste apart, according to the late Eric Eweson, who spoke at the first *Acres U.S.A.* Conference. Even without scientific intervention, conditions create them. It has been said that bacteria that produces food for corn are different than bacteria that produces food for soybeans. How microorganisms could produce for corn not for soybeans without an ensuing war seems elusive. This does not appear to be the case in the soil. In some cases farmers have sprayed on bacteria with good results. I attribute this to the fact that they have brought in a new division of troops. If this new division starts to slaughter the old troops, everyone loses. Wherever there are bodies, this stimulates activity in the soil for rebuilding its structure and producing energy so crops can grow.

Siegfried Luebke of Austria buys his bacteria by genus and species from a biological house in Maryland. He mutates them so they can operate in the soil or in the compost pile, whatever. By the time he is finished, he has a new microbial critter. The microorganisms he buys have to be kept at a certain temperature or they die. His finished product can survive the hardest conditions, drought, heat, cold. It is safe to say that these microorganisms came out of the soil in the first place. The laboratory made them a sissy-type bacteria. He in turn toughened them up again so they could go back to work and do the job. In healthy soil, this same phenomena will occur naturally. If the toxic levels are so great that bacteria are unable to protect or produce, then they are not going to be functional. It may take several hundred years—if the toxicity levels are too high—for bacteria to resume their roles effectively.

To understand this, all we have to do is look at nuclear waste. In the case of Siegfried Luebke, as reported in *Acres U.S.A.*, his soils did not put cesium into root crops at the time of the Chernobyl accident. It must be theorized that his microorganisms were strong enough to take cesium apart, preferring instead to have potassium engineered and delivered to the plant roots on signal because the plant can determine what it wants. Soil carbon was acting as the police force. In the scheme of things, the plant determines what it needs and then releases enzymes for bacteria to deliver. Foreign bacteria are often brought into the equation. The damage they can account for can be irreparable. It is very much like taking two hives of bees and merging them together. Say one hive doesn't have a queen and the other one does, and you are trying to salvage the bees for whatever their value is. You would take the top off one hive, put a piece of newspaper on it and place the hive that you want to introduce on top. When the bees eat through that newspaper, they will be buddies and friends, and they will co-mingle and become a socialized section of that existing strong hive under the control of the queen. The same situation applies when we are talking about bacteria. They should be introduced into an area where there are no bacteria, but if these bacteria are placed in an area that is flourishing or an area that is starting to flourish, there wouldn't be any point in doing that. Carbon's role is never far away when we consider microorganisms, aerobes and anaerobes. Let's say that a unit of carbon is capable of absorbing and holding five units of toxicity, but the unit can only absorb three. Two are loosely attached. These toxins can then be taken off of this carbon unit with relative ease. It can be broken off as an electrical impulse which forces bacteria to do something with it. It is now a floating toxin and has to be put someplace. As it falls off into the middle of the colony—where there is no surrounding structure for them to attach it—serious problems emerge because microorganisms have no transportation system. But if another unit of carbon enters, and it has a capacity of three and we have five to be absorbed, we now have 2.5 units of toxicity per unit of carbon. This allows bacteria to function in a more normal pattern. Carbon, again, is very important in this process. If the amount of toxicity retained in this carbon structure is of such a character that bacteria cannot get rid of it, they will do nothing with that. It will remain dormant. If it is a heavy enough form of energy, and that unit of energy has enough moisture and causes a connection between it and the magnetic field, as the earth swings around and vibrates those materials will sink lower into the substrata. As it sinks, once it gets beyond the zone of the bacteria, and they have no more concern with it. So bacteria have a most important function in the soil for life to continue. Without them, we would perish from this planet. Time is marching on. We must act now!

7

What Plants Grow Where

While I was in Florida counseling a group of farmers on the paucity of carbon in their carbonless soils, one of my clients handed me an Okeechobee gourd. Laboratory scientists had already concluded that a special resistance to powdery mildew was locked in its genetic code. Moreover, that nemesis of many Florida growers, tobacco ringspot virus, couldn't make a dent in the Okeechobee gourd, nor could bean yellow mosaic or cucumber mosaic. Someone had the information that this special invention of the Creator was loaded with linoleic acid and had a superbly high load of bitter cucurbitacins, a natural attractant for corn rootworms and cucumber beetles, products of interest to integrated pest managers. The question posed that session was simply, Why? Why do plants grow and thrive where they do, and also, *How?*

I had few answers then, and have only a few now. I recall George Washington Carver's admonition when confronted by such a barrage of queries. He said something to the effect of, *My, my, you want to know too much. Your mind is not big enough to understand all of creation, so why don't we settle for knowing a little bit.* And with that he started taking the lowly peanut apart. His probes into the unknown represent one of the few cases in scientific history that fragmentation has been used to understand the bigger picture. Most investigators see their own limited field as the governing factor for the whole.

There are many varieties of soils. Some have a large inventory of clay particles and a high capacity for energy exchange. These are fertile soils, in short. Some soils have no freedom of energy exchange. Poverty soils would be a good term to apply to such cases. And between the extremes are all the possibilities that calibration of fertility can account for. Finally, there are soils that are not only devoid of energy exchange capacity, they actually suck up energy from other sources and lock it up. We may never know all the answers as to what makes plants select environments, migrate, proliferate or die out. If we considered all the factors involved, we would be lost in endless speculation and computation. Here the kinetic theory of gases may help us to understand. Any gas is composed of free molecules in constant motion. These molecules travel at different speeds, they drive in all directions, collide with each other, and impact on the walls of a cylinder holding them. Pressure is nothing more than the measure of impacts, a calibration of energy molecules striking the walls of the container. Not even the biggest computer could calculate the path of each molecule.

"If by ill luck I happened to know the laws which govern them I should be helpless," said Jules Henri Poincare. "I should be lost in endless calculations and could never supply you with an answer to your questions. Fortunately for both of us, I am completely ignorant about the matter. I can therefore supply you with an answer at once. They may seem odd; but there is something odder still, namely that my answer will be right."

I doubt that many field agronomists will be able to answer many of the esoteric questions about the Okeechobee gourd, or lemons, or even wheat, corn and alfalfa. And I readily confess that I am ignorant of many of the laws of nature. This, indeed, may seem odd, as Poincare said, but odder still will be the fact that these few observations on carbon, energy and plant growth are right.

Plants grow as they do where they do because energy fields

that are necessary for their growth are in place and functioning—to some degree. When weed seeds and grass seeds increase and multiply, they obey that biblical injunction because the energy fields necessary to break down seed coatings, to penetrate, to get moisture into the seeds are alive and operative, so to speak. Many scientists belive that plants grow as they do where they do because of deficiencies in the soil. This may be an unsatisfactory explanation. It might be more correct to say that plants are regulated by toxins as well as nutrients available to them.

I have seen Highway Departments strip away all the topsoil in a road area. For a good twenty-five years, nothing would grow on such terrain. Finally the first cover to grow would be some of the grasses, chiefly very rough and rugged grasses. These plants would fall down in time. The sun could not penetrate the surface areas involved. Nor could air. Moisture under such cover was trapped. If you get a good crop of foxtail, I guarantee that the soil will be hard to dry out almost as if a plastic tarpaulin had been pulled over it. Trapped moisture allows bacteria to grow. These are chiefly anaerobic bacteria, and they start to process organic matter to humus. Once this happens, weeds appear as if by magic—brambles, briar type bushes, soft woods, elderberry, and so on. The cycle will continue in the presence of enough moisture until the hardest woods arrive, namely maple, locust and oak. Each will grow in the time frame nature decrees as proper. There is very little man can do to alter the cycle. The same general principle applies to farm crops. You can't very well plant a corn crop in the forest unless you take the trees out, even if you water the plants. By way of contrast, marijuana will grow under tree canopies in very poor soil. It grows there because it has the right conditions. It likes shade and wherever there is shade, there is moisture, and the marijuana plant can take it out of the air much as all other plants and trees do. In the germination of seeds, biogenic carbon comes into play. If a soil is not contaminated by chemical toxins, then the loss of humus causing excess CO_2 is not possible. As noted in *An Acres U.S.A. Primer*, foxtail best germinates in the presence of excess carbon dioxide and the absence of oxygen. Compaction often allows the decay process to trap carbon dioxide that ought to escape. Everything comes into play as the plant growth drama unfolds, including the flow of magnetism from the equator to the poles and energy from the cosmos to soil and plants.

If the foxtail plant can grow eight or nine feet tall, the highest concentration of chemicals will have leached down because of the water run off, and if the soil is hard and sealed, anaerobic microorganisms will convert toxins into gases, but the gases cannot escape. The plant now becomes the escape route through its root system. The only way gases can get out is to germinate a special seed that is otherwise dormant. The process activates the hormone and enzyme systems of the foxtail and in effect says, *This is the year for you to establish the kingdom and grow.* Where there are no weeds, there is a greater capacity for development of toxic conditions in foodstuffs. If no weeds or grasses are allowed to grow, toxic conditions are exacerbated. Thus it appears that a complex of weeds and grasses is critical to healthy soybean, corn, and wheat production. Once the chemical toxins go into decline, then weeds and grasses start to diminish. This makes increased productivity possible because good energy is overriding bad energy, toxic energy being considered bad. But when the good outweighs the bad, an increase in quality is possible. If yields are forced, imbalances are worsened and the nutrient structure is worsened.

There is an old adage that weeds index deficiencies in the soil. I have tested may weeds, and I have taken soil those roots were growing in. Using standard testing procedures, I have not been able to construct an index for deficiency in the soil and the structure of a plant. This suggests that a full explanation must invoke an added consideration. I suggest that it is the function of weeds and grasses to act as purifying agents. These agents do not act like carbon, which holds toxins. The weeds and the grasses grow on toxins and disperse them. That is why it is a terrible mistake to cope with weeds by dumping poisons on them. To deal with weeds, a farmer is pretty much restricted to some sort of clean tillage program. Even at that, it becomes necessary to use between the row cultivation, always allowing weeds that seek to grow within the row itself to grow. This way the farmer can be relatively sure to have a toxic free product growing on soil that is in balance nutritionally. In almost any field, there will be two minor weeds and a predominant weed. The same thing applies to grasses. Usually a minor weed this year will disappear the next year, and a major weed this year will then become the minor weed going out. The minor weed coming in now becomes the major weed, and a new weed or grass comes in to replace the minor one. It also will be observed that a weed or grass will come back in another season, but in much smaller numbers and a smaller size. As these weeds go through their cycle, on the incoming side, seeds will exhibit a certain density and structure. Much the same is true on the outgoing side. This appears to be nature's way of placing into the soil weed seeds that will dominate in the future. Weed seeds can lay dormant for hundreds of years, maybe even thousands of years. Crop seeds are not so easily preserved in the soil.

I have made this observation. During a year the ragweed was going out, it would come up, say, twelve inches, and then it would form a crown. Out of this crown would come a series of seed protrusions. It would be an odd looking thing. We can visualize what ragweed looks like. It is usually a bushy type weed depending on how much energy it had to grow, but in this case I was looking at something that came up and then stopped growing. Then, all of a sudden, it mutated the growing cycle. Visualize, if you will, a crown with perhaps ten or fifteen individual little stems all loaded with seeds, and then the next year they were not there. The seeds were not perfect that next year, most being malformed.

I think we have been listening to the universities and the chemical people too long, and we have not been looking at what nature really does and how nature functions. It is true, nature functions pretty much by establishing the terrain for a certain type of a crop. This is why it is always so difficult to grow a crop outside of its natural environment. We cannot grow bananas in southern Canada very easily. And sugar cane does not belong in Colorado. There are optimum places in terms of elevations, in terms of rainfall, in terms of temperatures and soil temperature. These things undoubtedly figure on a micro-scale. It must be observed that weeds and grasses grow in a certain environment, at a certain elevation, in a certain rainfall pattern, under certain tillage conditions. But with chemicals of organic synthesis put on production, a new factor is introduced. This factor frequently engineers a better environment for the weed to grow in.

Weeds in their environment are very much like bacteria. They have the ability to change their structure over time to adjust to the environment and to correct toxic imbalances that may exist. They do this in rapid order. In some cases they set up a rhythm between one weed dominating one year, another dominating the next. The entire process might very well be governed by how the spring broke, how high temperatures came. These several factors cause changes to take place in bacterial proliferations to speed the pace or slow it. Rye provides a good lens through which to observe a negative energy growing grass. We hear much about how it controls the weeds. It controls weeds by soaking up all the energy weeds normally grow on, causing everything else to starve to death. This is basically what happens in nature. One plant does not grow on all of the energy fields. Each grows on a particular group of energy fields. Ragweed and lambsquarters grow on widely different energy units. Pigweed grows on still another complex energy system. Unfortunately, we have not yet learned how to measure these positive and negative energy fields, although I believe I am close. In a real sense, I think we are very much like an infant freshly emerged from the womb in what can be done and what is really known. Many people using the low frequency energy scanners view radionics as a probe into unknown energy systems. They know the shortfalls. They also know that people who haven't done a hundred hours of work on the instruments are not worth talking to.

Weeds are only marginally valuable in establishing what the shortages are in a soil system. They will not tell, for instance, whether the phosphate load is correct. If there is a phosphate shortage, the plant growing there will not add to it. The deficiency will still remain. The plant isn't going to change the equation at all because the phosphate has nothing to do with it. The weed would use up the toxins present, thereby allowing nutrients to be released from rock storage. As carbon is converted, toxins held are released. By causing energy to be released, they cause weeds and grasses to grow.

We know that the carbon is a police force in the soil. This police force will allow weeds to take up certain energy fields. It is almost as if the police force said, *"You" live in this house, and "you" live in this house.* Carbon not only protects which one goes into an area, it allows entry only to the plant that is *supposed* to go there at the same time. Carbon plays an important part in determining the flow of energy—how it is used and how it is converted, how it is consumed and how it is recycled.

8

Siegfried Marian's Insight

Some twenty years ago, I read a wonderful little book called *Bread from Stones.* Its author, Julius Hensel, was a physician, a critic of Justus von Leibig's salt fertilizers, and a towering refutation of allopathic medicine. His health findings carried such titles a *Macrobiotic* and *Das Leben,* all vintage 1880s era material, all more modern than the best the American Medical Association and modern agronomy have to offer.

He attributed "the futility of orthodox medicine to a lack of knowledge concerning the role of the physiological minerals, which give electric cohesion and relative durability to otherwise unstable materials in the body." The quotation marks are used here because a part of the above sentence came straight out of *Soil Magazine,* October 1950, editor Siegfried Marian, who pronounced Hensel on the right track, and this prompted me to reread *Bread from Stones* again.

Hensel saw a direct connection between electrical cohesion in the human body and soil in agriculture. Certainly, he was one of the first to create a stir by downgrading and condemning the chemical manures then sweeping Europe and taking command of all centers of learning. He was almost entirely alone when he recommended stone meal for application to the field. By stone meal he meant crushed stones that had been fired before crushing to achieve a powder-like consistency. He favored ground porphyry, an igneous stone, and he made a point of relating quality food with the presence of soil containing full loads of all the important minerals. He didn't talk about trace minerals because they hadn't been identified by science to any extent, and yet he seemed to postulate as much because complete pulverization was indicated.

But it was the scientist Marian who caught my attention early in the game, not so much because of what Hensel said, but because he picked up where Sir Albert Howard left off.

I had been experimenting with various coals and formulating NutriCarb for some time when a client mentioned Siegfried Marian, who even now is no household word among farmers and scientific agronomists. Sheer persistence and several years have earned me a small file of his papers.

For a period of no more than three years, he published a small magazine with no commercial advertisement, and then he passed away. The notes that have come my way are few yet some idea about the scope of his investigations emerges from a notebook entry on cloudy charcoal—No. 5062. This means that between 1938 and late 1950, he tried 5,061 different processes before he was satisfied with the results. He wanted a type of charcoal that would not be gritty, which would have the greatest possible reactive surface, and which would be impregnated with a molecular film of a potent vitamin. He wanted a charcoal vitamin, and he got it. This charcoal vitamin was developed to prevent diseases from toxic causes.

Over the years I assembled a few of the copies of *Soil Magazine*—but not all. These treasures simply vanished off the face of the earth with his death. SM took on an esoteric meaning—it stood for Siegfried Marian and it also stood for *Soil Magazine*. Many of the lessons contained in Marian's work have surfaced in the several chapters that fill out the pages of this book. I have mentioned soil viruses. Glancing over the back issues of *Soil Magazine*, I found—in addition to the fact that research has shown the virus to be the product of enlarged and swollen proteins—the point that the virus is life's enemy. It is "just a protein which has lost its freedom of energy exchange and is thus sucking energy from still energized proteins of living substance." In *The End of the Virus* [*Soil Magazine*, May 1950] this simple idea is traced through all its complex details.

"The presence in soils of suitably built carbon-oxygen compounds is therefore a safeguard against the formation of virus, and plants can never acquire virus either by generation on their own, or by infection. In the case of infection from outside, the presence in the plant sap of soluble carbon-oxygen compounds, taken up from the soil, destroys virus in the following way. . . The virus structure is in the main a tangled web of electric fields which are controlled by a positive electric charge. The presence of a negative field, as offered by the oxygen in carbon-oxygen compounds, removes the primary positive electric field of virus particles and as soon as this happens the virus breaks up into its components and ceases to exist. All these proteins which were previously held inside the virus structure are set free and are carried away singly."

In August of 1950, Marian went on to explain his concept of actumus, much of which has been discussed in earlier chapters. One of his passages is worth quoting: "The protecting influence of actumus is given by its capacity to hold a percentage of electrically charged oxygen. The formation of virus is therefore prevented at the root. The vital part played by actumus in the *natural* nitrogen cycle of plant growth [makes] all the pretexts for resort to poisonous artificials for this purpose totally unnecessary."

As I read on, it seemed to me that some of these actors continued to return like a strayed penny. They read their lines like well-paid characters in a play.

Life is organic and built on carbon. Chemical fertilizers destroying the carbon in the soil also destroy the possibility of the formation of hormones and vitamins which plants store in their reproductive parts. This is the capital of life which must diminish from one generation to another if the carbon in the soil is destroyed by chemical salts. I listened with thirsty ear to

VIRUS

Under this title The Cambridge University Press published a book by the eminent expert on virus, Kenneth M. Smith, F.R.S. It is remarkable that in spite of difficult terms on very scientific lines, the author was able to give an account which is written in a language so easily understandable even to the unschooled biologist, farmer and even layman. . . But the author has missed the main problem of virus, i.e., its connection with the soil in which plants grow. We cannot rightly say that he has missed it, because those test tube researchers who are divorced from the soil have not even made a start to try for a link between virus and soil. We must again refer to the outstanding personality of the late Sir Albert Howard who has shown us that cattle raised on soil with humus are immune against foot and mouth disease, which is a virus disease. I shall always remember his words when he said that his cattle could rub noses with infected animals without contracting the disease. Professional research workers took no notice of Sir Albert's findings, because they have a perverted sense of scientific truth. They think that all help and relief can only come from an elaborate chemical process, the more complicated the better, and the more artificial the better too. Then what they have found in such ways must be produced in vast factories costing many millions of pounds.

I have devoted much thought and work to prove that Sir Albert was right, and I shall report about it later. It is not much good to know all about virus, to photograph it by the use of the electron-microscope, if we are not given a method by which to eliminate the immense danger and crippling economical losses to humanity.

Look at the tragedy in West Africa, where a few million cacao trees are felled in order to combat the virus of the swollen shoot disease. Look at the slaughter of cattle because of foot and mouth disease, just to mention these instances as examples. The relief from virus can only come via the soil. We must go down to the soil if we want to find relief from the greatest danger to life on this planet.—Siegfried Marian everything, including that matchless body of literature known as the *Acres U.S.A.* tapes, always hungry for more. It seemed inevitable that my odyssey would take me back to Hungary, at least in spirit. Actually I had to go no further than the pages of *Acres U.S.A.*, which at one time carried the works of the late Francisco R. Tamas, then of the Dominican Republic. Actually Tamas was born in Hungary. He did not migrate to the Dominican Republic until 1931. Before that he spent some forty years researching non-salt fertilizer practices in Europe, Asia and South America. He graduated from the Royal Hungarian Polytechnique University, and like Siegfried Marian, he studied in Vienna. He also studied in Paris and Cambridge, and finally achieved international standing when he was elected a member of the International Association for Quality Research on Food Plants in Helsinki.

In *Health Food Can Only Come from Healthy Soils*, Tamas only added to what those fragmentary Siegfried Marian papers had already made a matter of record. A few lines can be quoted to illustrate the point:

"We wish to emphasize that healthy plants are the vital basis for men's health.

"To mention a scientific proof for this statement we remember that Dr. Professor A. Voisin, member of the Academy of Sciences of France, conducted investigations which discovered among other things that the functioning of the thyroid gland is under the strict control of the elements of soil. We do have therefore the certainty that through the thyroid gland soil exerts a determined influence over the hyprofiso-suprarenal (kidney) which occupies a fundamental place in the defense mechanism of our organism against all external attack!

"Professor Hans Selye, of the Montreal University—creator of the concept of *stress*—demonstrated that in the defense against this abnormal state, the hormones play an important role over the hyprofiso-suprarenal by which they are capable to modify the resistance of the tissues against the pathogenic agents.

"It happens that once we have *reconstructed* the soil into a healthy soil in which healthy plants can thrive, our organism receives through the plants we eat all the necessary trace elements, vitamins, proteins, enzymes etc., which this soil reconstructor supplied to the soil and by eating poison free vegetables and fruits our organism slowly but steadily detoxicates itself and all reason for sicknesses gradually disappear."

Everyone has a term. I have my own NutriCarb. Tamas had his Soil Health Reconstructor, which, he said, "was wholly complete in itself, releasing and making available by its action all growth and nutrient properties already latent and plentiful in soil, so that no crude chemical salt need be added. This soil reconstructor—scientifically devised—contains all the elements of which our earth is composed, therefore it factually recreates soils it was when our planet was a *paradise* because its soils were not yet abused, depleted and degraded as they are today. What happens to the soil once treated by this reconstructor is that it gets back all the main enzymes, auxins, auxons and nearly all the trace elements in a naturally balanced formula, which makes the soil biologically balanced and healthy."

I do not know whether Tamas ever met Marian. It is very likely that the former and the latter met in spirit, and perhaps on paper. They were contemporaries, even though Marian died in what we consider mid-life, and Tamas lived to a ripe old age. Even though both men have been ignored by what Sir Albert Howard called "the republics of learning," the time has come for the idea contained in their work. I believe I have tapped that virgin knowledge and given expression to it in the product called NutriCarb.

Having said this much, we can now summarize, abstracting and extracting from one of those rare journals published by Marian for hardly three years.

Every plant [he wrote] *has its complete schedule of work and requirements. Exactly as in the manufacture of a technical article we cannot make the article if we are given too many nuts and insufficient bolts. The work going on in a plant is much more of a precision job than the manufacture of a car. The flow of elements must be sorted out and it must be exactly controlled. If the work of building up the plant is to succeed, the elements must be supplied at the correct rate and in the correct mixture. There may be unwanted elements which by their great numbers and by their strong electric forces may exclude and subdue other elements which are wanted by the plant. Many young seedlings die because they were put into an environment where forceful elements like iron have come into the rootcells in great numbers. Owing to their strong electric forces they resist the pull through the vessels of the plant. They settle in the roots, block all passages and force a closing down of further work in this plant.*

To prevent such anarchy in the movements of elements, life establishes its police in the soil. This police is not a strange force. The police of life exists in every living thing and it is organized during the life of every living thing and posted on all important points. This police consists of extremely reactive carbon crystals in a hexagonal shape. They control the movements of elements inside every living thing. Unwanted elements cannot do what they like. The cannot in the presence of carbon crystals force their way through the factories of life and break up normal work. Unwanted elements are caught by the police and arrested for good. Should a living thing require a little of the unwanted elements at one time or another, the police will release as much as is wanted at a certain time. Anarchy is not allowed in the organization of life.

No living thing lives forever. When the work is finished, the living thing has done its functions and serves other purposes. It either dies the natural way, or it is killed by force whence it serves as a raw material for building other forms of life in the form of food. The living thing which dies naturally is also subject to being used as a raw material for building other forms of life. Bacteria feed on living things which die naturally.

Whatever the ways are through which a living thing goes after death, it is eventually completely dissolved into the elements from which it was built. None of the numerous patterns of compounds of carbon with other elements remain. Since most of these compounds were made from water and carbon dioxide gas with the help of sunlight, a great bulk of dead matter will leave nothing but a very tiny heap of solid matter when its dissolution is completed.

This tiny heap is of very great importance. It contains the police force of hexagonal carbon plus those elements which the living plant has extracted from "dead" soil by the respiration of its rootcells. The elements go back and so does the police. The loss of the elements would not be so tragic. Our earth is so full of them that the requirements of life can be met for times which cannot be calculated. But, the loss of the police has catastrophic sequences. The police of life is not so abundant in normal life processes. The police exists only in proportion to existing life. If a part of existing life is handled wrongly so that its police is not returned to the soil, an equal amount of life is lost forever, because without the police from previous life new life cannot exist.

On the borders of life in the soil, where tiny rootlets pump water and dissolved elements strictly controlled by the police force of life, there are other living things which are also employed as "sub-contractors" for the provision of an element which does not come from soil. This element is nitrogen. It is a gas and exists in the air. Unfortunately nitrogen is not so obliging as oxygen. Nitrogen is a lazy element. Plants and all living things need it for the manufacture of proteins. It must be got by all means. Bacteria, fungi and molds are engaged for this purpose. These living things are specially adapted to trap nitrogen from the air and break its laziness by forcing it into chemical compounds which are not so lazy. This work is being done on a barter system. Plants give these nitrogen workers a bit of sugar of their own manufacture, and in exchange for it these organisms trap and prepare nitrogen compounds which plants require in their factory for the complete working process.

These organisms which are of such great help to other life are in danger of being ousted by bad boys in the world of bacteria, fungi and molds. There are more bad boys in this society than good boys. The bad boys like to take sugar from plants and do nothing for it. They sap life and spoil it. But, the bad

CLOUDY CHARCOAL

To ease the bombardment of questions which come in about Cloudy Charcoal, these lines are offered to those who have taken this preparation with very great success. I refute the suggestion that there is something magical in the preparation. There is rather more hard work and hard thinking in it.

If you want to understand the complex which is forming charcoal, you must first know that charcoal is not a fixed substance. From one and the same piece of wood or other organic matter, I am capable of making a few hundred types of charcoal which differ from each other. The properties of a piece of charcoal depend in the first instance on the raw material, but once the raw material has been selected, the way and the method by which the raw material is subjected to the distillation process decided the properties of the product. To explain this in a short article is quite impossible.

The distillation process of organic matter, leading to charcoal, is greatly influenced by a multitude of factors, of which the most important are: temperature, air inlet, injection of gases and vapors, and speed of the process. Variations between these factors and other factors not mentioned here, account for the multitude of different products which can be obtained from one and the same raw material.

In the realm of charcoal and carbon generally, I play a very lone hand. Because so few people know much about this complex, they cannot understand my reasoning. The difficulties are enhanced by the fact that all those processes which go on inside a charcoal distillation retort are of short duration, too short to subject them to the usual analytical examination.—Siegfried Marian boys are kept in check by the same police that control the flow of elements. Where the police of life exist, the life collaborators in form of bacteria, fungi and molds can supply undisturbed suitable nitrogen to plants and enjoy the sugar they get in return.

Where the police are lacking on the borders of life, complete anarchy is the sequence. Not only does the entry of elements become dangerous, but several bad boys from the society of bacteria, fungi and molds also proceed unchecked to enter and establish themselves in plants, animals and man. They are not satisfied with the bit of sugar which plant roots hold out for nitrogen workers, but they proceed to much richer stores of sugar and other delicacies inside living things.

Life is built from raw materials which come partly from the soil and partly from the air. The greatest part of bulk is water, which comes from the soil and from the air; carbon, which comes from the air; oxygen, which comes partly from the air and partly from the soil; and nitrogen, which comes from the air via the soil. The manufacture of living things takes place in a very well defined way from the necessary elements, and *the flow of raw materials is controlled by a police of hexagonal carbon crystals on the borders of life in the soil and inside every living thing.*

As in every manufacturing process of a chemical nature, there are waste products, effluents, etc., which must be removed if the manufacturing process is to go on. If waste products could not be removed, blocking of further work would be inevitable. Here again the police of life do their duty. If for one reason or another waste products from factories of life cannot be ejected, the police remove them from the passages and keep the ways of communication open.

What then is disease? Disease is anarchy on the borders of life caused by lack of police in sufficient strength on these borders of life. All diseases in plants, animals and man have their starting point on the borders of life, where rootlets attack non-living matter for raw materials. Let us be sure about the fact that life is a fight, and that the creation of life is not brought about by peaceful means. Life mines the soil and the air for raw materials for its existence. Life is on the offensive against the non-living, and its attacking force is protected by the carbon police. Where this police is weak or absent, life loses the battle. Instead of life going into the territory of the non-living for raw materials, the non-living invade life, and the forces of the non-living are anarchic.

The forces of the non-living attacking life in the form of diseases have "fellow travellers." These are the bacteria, fungi and molds and viruses, which do not help legitimate life on a give and take basis, but march with the forces of anarchy which are non-living and also attack and destroy legitimate life. Our doctors call such "fellow travellers" pathogenic bacteria, fungi, and molds. The word pathogenic means that they cause pain and disease.

The security of life for our human race is situated on the borders of life where plant rootlets attack the non-living. If our plants are successful in building up life free from disease, they will contain quantities of carbon-police, which—when plants are consumed as food by higher animals and man—will protect the latter from the invading forces of anarchy with its diseases. In this sense life becomes susceptible to disease when the borders of life are unprotected, and life is immune to disease if the borders of life are protected.

I mentioned the "fellow travellers" microorganisms which, though living, march with the anarchic forces of the non-living, causing diseases in those living things which belong to the sphere of interests of our human race. What about the virus? A virus, again, is nothing else than an anarchic attack on the living. Owing to lack of police, unwanted chemical elements armed with strong electric forces enter the living thing, and requisitioning raw materials from the living. They upset the work and building process of the living. A virus is caused by the burglars of life. These burglars are chemical elements which come in where they are not wanted. Into their sacks they fill materials of life and rob life. A virus is thus a chemical element with a sack of robbed materials. Supposing that in our social life burglars would have a free hand in the absence *of any police*, can we see how normal life would not be possible? The social structure of our life would become diseased and it would die. The same happens with regard to virus diseases in living things. These living things become robbed and die by the forces of anarchy of the non-living.

The human kingdom of life with its animals, plants, collaborating bacteria, fungi and molds is the youngest kingdom on this planet. The history of this human kingdom is very short compared with the history of our planet and with the history of the universe. The human kingdom in relation to the history of the world is a very tiny seedling which is fighting for its existence on the borders of life in the soil.

NutriCarb, indeed, is a policeman and performs all the roles involved in that vital function.

COAL TAR DRUGS

Hand in hand with this endless pouring out of drugs from coal-tar from our factories went another development which came from a different direction. This was study of diet. Scurvy was a disease which has always baffled doctors, and they were always conscious of the fact that this disease had something to do with diet. When Cartier touched Newfoundland in 1535 he found that the Indians living there had a cure for Scurvy, which was prepared from needles of conifers. This was 200 years before Europe had a cure for this disease which was very dangerous in ships or barracks, where diet was more confined to preserved foods. It was in 1720 that Austrian Army Doctor Kramer, showed the world that a little lime or orange juice, soon cured the disease, and in 1747 Huxham ordered that the British Fleet had to take vegetables on board. These facts initiated the study of what we now know as vitamins. These are substances which are contained in food of good quality, and without which the animal and human body cannot exist in a healthy condition. I do not propose to go deeper into the history and development of the vitamin theory, because it is now sufficiently well known that without vitamins we cannot exist. The fact also stands out that by keeping out diet with food containing vitamins many diseases due to vitamin deficiency, have practically disappeared. Credit for this development of medicine must be given to the doctors who have in this case freed themselves from the laboratories of the chemical industry, although this industry in now trying to push across artificial vitamins in their never ending appetite for dominating the drug market.—Siegfried Marian

9

The Carbon Connection

During the last twenty years, a considerable body of literature has surfaced to explain biologically correct farming. Except for the works of the late William A. Albrecht, which have been quietly ignored by the Soils Department he once headed at the University of Missouri, many of these works have reflected—as C.J. Fenzau once put it—"the beginning of an intellectual attitude in conflict with both agronomic science and industry . . . This virgin knowledge is finally emerging as a test and challenge to the introverted and integrated industrial exploiters, and absconders of our most precious resource."

That intellectual attitude, fathered by Bill Albrecht, Ehrenfried Pfeiffer, Joe Cocannouer and many others, is now gaining a hearing. The Albrecht papers have captured the nature of the challenge, of course, but some of the Albrecht papers are not now available in book form. In preparing the text of this book, I was handed several taped sessions the *Acres U.S.A.* editor, Charles Walters Jr., had with Albrecht during the last year before the old gentleman departed the scene. Attached were data from *Variable Levels of Biological Activity in Sanford Field After Fifty Years of Treatment*, a paper Albrecht had written, together with several technical bulletins. What Albrecht sought to make clear—as a basic lesson—was the carbon-nitrogen connection, and what all this had to do with humus formation and the survival of organic matter in the soil. The composite became a lesson in basics, for it is a fact that with every report a farmer reads, he is compelled first of all to return to the basics. The following appeared as *The Carbon Connection* in the September 1985 issue of *Acres U.S.A.* It furnishes us a focal point from which to *review* the full scope of the carbon connection this book provides.

Carbon (symbol C, atomic weight 12.011) is an important chemical non-metallic element. Carbon takes all three forms of matter—solid, liquid and gas. In its purified state, carbon represents the world's most treasured gem—the natural diamond. Charcoal, soft and hard coal are largely carbon. The purer the carbon the better the quality of coal. Petroleum, oil and gases pumped from the earth are high in carbon content. Carbon is also present (as carbonates) in the composition of many minerals, such as chalk, calcite, dolomite, calamine, and others.

However, every form of carbon found in our earth has its origin from former plant or animal life, the residues of which had somehow become fixed in our earth's surface or crust. The ore-bearing rocks of the earth originally contained no carbon, but it had to be acquired from the atmosphere as carbon dioxide principally through photosynthesis, whereby plants converted it to living tissue.

Under certain circumstances carbon is an inactive element. A piece of charcoal, graphite or diamond may lie unchanged for ages in ordinary water. All will be without taste or odor and cannot be melted. Under other circumstances carbon may be very active and mobile as witnessed by the fact that it may enter into more than 20,000 [when Albrecht wrote, 20,000, now 3,000,000] known chemical combinations. In fact, the entire field of organic chemistry may be aptly referred to as carbon chemistry.

Throughout every phase of agriculture, carbon is the indispensable element that supports and enervates every function of microbial, plant and animal life. Carbon is the great biotic, organic, biochemical fuel that gives heat, speeds both inorganic and organic reactions, but inorganically it deals with living matter—and it is the living matter wherein it has its major role. On a dry weight basis, it represents upwards of 50% of the total composition of all plant growth. Without carbon there would be no plants, and without plants there would be no life. It is the basic element of our life sustaining universe.

Every form of animate life, which in turn depends upon the products of plant life for subsistence, is basically dependent upon the element carbon. Carbon enters into the composition of all plant and animal tissue. Furthermore, the element carbon is nature's cornerstone upon which all elements of nutrition are built. Without carbon there can be no protein. Without carbon there would be no fats or oils. And without carbon there would be no vitamins, amino acids, enzymes, or hormones. So closely is carbon linked to all of the life processes and functions that without it life could not exist.

While carbon may enter into thousands of different chemical compositions, agriculturally speaking, two forms are worthy of consideration. The first are the more purified and stable forms as found in the humus constituents of soils and in the composition of proteins, sugars, starches, oils, fats. The second is one of its gaseous forms, i.e., carbon dioxide.

Carbon dioxide, or carbonic acid gas, is a well known compound containing one atom of carbon and two atoms of oxygen. It is colorless. It exists in small quantities in atmospheric air at the rate of only three or four parts in every ten thousand. It is formed by burning of fuel, by the breathing of animals and humans, by fermentation, and by the decomposition of animal and vegetable matter. Carbon dioxide is incombustible. It will put out the flame of a lighted candle, gas jet, or even a piece of phosphorus, when these are placed in a jar filled with this gas. It is not poisonous. However, too much carbon dioxide in the air prevents the lungs from obtaining needed oxygen.

Carbon dioxide moves into plants, principally, through the

stomates (breathing pores) of their leaves and is converted to form a large part of the tissues of the vegetative world. The starch of wheat, oats, rice, barley, potatoes, bananas—in short, all of starchy food plants—is built up of carbon dioxide and water. The gas (CO_2) enters the leaves, and water moves into the roots. A farmer who sells hogs, cattle, butter, cheese, sugar beets, flour, potatoes, or any other kind of produce, is marketing little but water and gas, millions and billions of pounds of which will again become dissolved in the atmosphere. The end products of all complete digestion are again water and carbon dioxide.

Through the miracle of photosynthesis in plants, energy from sunshine, carbon dioxide and water are combined to form carbohydrates which supply heat and energy for the support of life. To this combination must be added another element, nitrogen, which supplies the "growth factor." Nitrogen alone energizes and supports all of the functions of growth and development for every form and species of life. Whereas carbohydrates and fats supply the "go power" to life, nitrogen supplies the "grow power."

Nitrogen is an element existing in nature as a colorless, odorless, tasteless gas. It forms about 78% of the weight of the earth's atmosphere which, translated, means that there exists 39,000 tons of atmospheric nitrogen over every acre of land and sea on the earth's surface. Nitrogen is an essential constituent of all animal and plant tissue. Nutritionally, it is one of the main elements in amino acids that builds the protein molecules which provide and stimulate the growth factor in every living cell.

Nitrogen—like carbon—is an airborne element. Like carbon, it was not an original constituent of the rocks that form the crust of our earth's surface. Nitrogen—like carbon—had to become fixed in the soil principally from decomposing residues of previous life. It is noteworthy to realize that while nitrogen represents 78% of the atmosphere's composition (39,000 tons over every acre), carbon represents only 0.03% as carbon dioxide (7,698.2 pounds over every acre).

With the exception of legumes (in symbiosis with root nodule-producing bacteria to fix atmospheric nitrogen and several non-legumes in symbiosis with root-entwining fungi as a less able nitrogen fixing combination), plants depend upon deriving nitrogen from the soil. To accommodate this necessity, nature has devised a special nitrogen storage facility in the soil itself. This process, activated by millions of soil microbes, brings about a product of certain stability involving the elements carbon and nitrogen. In this process, all forms of organic matter—ranging widely in their carbon to nitrogen ratio (alfalfa hay, 12 to 1; rotted manure, 20 to 1; sugar cane trash, 50 to 1; cornstalks, 60 to 1; straw, 80 to 1; sawdust, 400 to 1, etc.) are reduced to a single standard formula of 10.4 parts carbon to one part nitrogen.

When organic matter has thus become reduced to the ultimate 10.4 parts carbon to one part nitrogen ratio, the end-products will be represented as brown colloidal bodies collectively known as "humus." The process whereby humus is formed is still as much a mystery as life itself. Someone has referred to humus as "nature's womb," where the principle of life has its beginning and from which it is eternally sustained.

Humus represents nature's one and only device for the storage and holding of both nitrogen and carbon in soils in certain natural ratios. Humus is insoluble in water and thus is not readily subject to leaching. Neither can either carbon or nitrogen be naturally leached out of the humus, though available to microbes. Nitrogen, fixed directly in the nodules of leguminous plants, is subject to the same process of reduction to humus as other organic matter. Only the soil microbes which contrive to fix and store nitrogen in the first place can reactivate it again to supply their wants and needs when it is in combination with carbon.

Every crop leaves behind in the soil a measure of organic matter represented in the root systems of the plants. This organic matter is in turn reduced to humus by the soil microorganisms. If this conversion did not take place, the soils would become clogged with such debris, much the same as do the drains in your bathroom. Through such conversion the soil's microorganisms, whose cell bodies have a narrower carbon to nitrogen ratio (5 to 1) than does the humus they perfect (10.4 to 1), have a positive need for a sufficient nitrogen supply.

In such manner nature has—throughout the ages—built and maintained her virgin soils to the utmost peak of balanced fertility. Although soils may differ widely in the kind and quantity of natural minerals necessary for best performance. Nature has provided special species for each area to accommodate the tolerances that such soils provide. Nature would not try to grow alfalfa on the soil where cranberries do best.

To accomplish these many miraculous works within the soil, nature depends solely upon the efforts of billions upon billions of microorganisms that swarm in the soil. So tiny are these organisms that 500,000,000 of them would weigh less than one gram, and there are four hundred fifty-three grams to a pound. These tiny organisms are ever alert, willing and frightfully voracious. Foremost and absolutely essential to their well-being and continued activity, is the element nitrogen.

In the same manner that a colony of honeybees will send its workers to collect nectar from thousands of flowers and then convert it into honeycomb as a perfectly processed product for its future food supply, just so do the various soil microbes process and store the products of their activity in a uniform product (humus)—also as a guarantee of future food supply.

In a mineral soil which by laboratory analysis reveals a 3% organic matter content in the top eight inches of soil, one may expect to find 60,000 pounds of such humus. Upon further analysis of this humus, it will be found that it is composed of approximately 34,800 pounds of carbon and 3,000 pounds of fixed nitrogen. Likewise, a soil containing only 1% organic matter would show only 20,000 pounds of humus with 11,600 pounds of carbon and 1,000 pounds of fixed nitrogen. The fertility level of any soil may be judged, generally, by its organic matter content.

When a farmer, after picking his corn crop, plows under the residues—stalks, leaves, husks, etc.—he will create a state of veritable mayhem in the soil. The sudden availability of such carbonaceous raw organic matter challenges the soil's microbial life into immediate activity. However, the wide carbon to nitrogen ratio (sixty to one in such material) poses the necessity for an extra nitrogen supply. This nitrogen is acquired by the microbes by simply seizing and appropriating the free nitrogen in the soil. Through such action, plant life may actually starve for want of nitrogen which the microbes are temporarily using to build their cell bodies to perform their function. It is well to realize that in all soils and plant relationships the soil's microbial life will eat at the first table, and plants must accommodate themselves with whatever is left over.

When a large mass of such highly carbonaceous organic material is placed in a soil which has a low organic matter content and a low total nitrogen level, a seriously unbalanced situation will prevail. Without a sufficient nitrogen supply, the soil's microbes cannot perform their intended tasks and much of the material will be left unconverted to be plowed up again during the next rotation. Such a situation is unfavorable for good plant growth and may be avoided if the farmer will simply provide one pound of available nitrogen for each one hundred pounds of such carbonaceous material as he turns under. Under such condition, enough nitrogen will generally be provided to satisfy the needs of the microbes and yet allow some for the plants.

When an excess of nitrogen is supplied to the soil (an amount over and above the requirements of the soil microbes to perform their various functions), a reverse situation will be effected. The excess nitrogen will be avidly seized and appropriated by the microbial life which, for lack of raw organic matter, will utilize the stored humus in the soil for their activities. Now, instead of seeking nitrogen, they seek carbon to provide the heat and energy for their functions.

When soil microbes are forces to appropriate humus in order to obtain carbon, they completely destroy the humus particle so that both the carbon and the fixed nitrogen are simultaneously released. In such manner an additional supply of nitrogen will be available—the nitrogen supplied as fertilizer and the nitrogen released by the soil. Recent research with tagged nitrogen revealed that more than 50% of the nitrogen utilized by the plant was natural nitrogen supplied directly by the soil.

It is this extra nitrogen, resulting from the destruction of the humus molecules in the soil, that gives the practice of heavy nitrogen fertilization its impetus. Soils that had a natural capacity to produce seventy-five bushels of corn per acre were suddenly found to produce one hundred fifty bushels per acre. The results were phenomenal and the increased use of nitrogen received the credit. This practice was considered as a major breakthrough for increased yields for all crops, and new nitrogen processing plants appeared all over the country like mushrooms in the spring.

Whereas farmers formerly strove to provide and maintain an adequate organic matter supply in their rotations, the present attitude has changed to where organic matter is considered unimportant because nitrogen fertilizers have proved such stunning results. This mistaken attitude is in error on two counts. It is expensive and it is highly depletionary. *Every pound of excess nitrogen applied to a soil will destroy one hundred pounds of stored humus.* The present attitude is responsible for converting modern agriculture into the world's greatest mining operation of its most basic and important asset—the humus stored in our soils.

Nature, somehow, prevents the accumulation and buildup of surplus free nitrogen in soils. Only by way of the carbon-nitrogen cycling processes can it become fixed and held in the soil. Generally speaking, any surplus nitrogen applied as fertilizer will disappear in the season during which it is applied. Nature has purposely designed soil to be bottomless to permit excessive nitrate nitrogen, and any other acids that may become toxic to plant growth, to conveniently leach from the root zone of soils.

This report is not intended to be an argument against the use of nitrogen fertilizers, but rather as an expose of the present day misuse of it. Nitrogen is an essential element for all plant growth. It prospers best when it is permitted to fully participate in nature's carbon-nitrogen cycles. This can be accomplished when a farmer will supply enough raw organic matter in his rotations to match his nitrogen application at a ratio of at least one hundred pounds dry organic matter to each one pound of nitrogen supplied.

For a decade or more, I have been using coal in soil systems. Some of the results have been measured in pragmatic ways: corn that can stand a killing frost, less corn borer damage, increased yields, drought resistance, increased fertility, increased earthworm populations, earlier maturity of crops, less disease, increased sugar content of fruit by 7%.

Hay fields alfalfa can be cut every twenty-five days with the same yields as thirty day cuttings. Application of three hundred pounds of coal has produced three times more hay on the first cutting with a four inch recovery in six days on alfalfa and eight inches growth on leaves of orchard grass. Cattle choose to feed where coal is applied on pastures. Feed grown on coal has higher nutrient levels and cattle consume less feed.

Rain arrives in abundance, but runoff remains at a minimum. The land speaks for itself and waves off the piece of paper bankers hold in their hands when there is profit in the system. Yet the role of carbon and its interaction with plant nutrients—a role so vital in preserving solvency for the family farm—is one of the least covered and understood by the farm press in general. The use of coal thus makes its own suggestion for farmers in a nation with soils that no longer contain the proper inventory of carbon.

"Healthy soils will pay off almost any sized mortgage within reason. Put carbon on your soil. The Creator provided you and the soil will provide you," is a quote I like because it is true, and not merely my quote.

GLOSSARY

Absolute temperature. The degree of the body in Centigrade plus 273.16 degrees. Absolute zero is -273.16 C.

Acid soils. Soils with a pH below 7.

Actumus. A process that in a sense is a mild form of carbonization equal to making carbon from wood without burning most of the matter to ash. The presence of Actumus crystallites in plant cells safeguards plants against the entry of toxic substances, viruses, malignant bacteria, etc. This is a term coined by Siegfried Marian that never made it into the dictionaries.

Aeration. Making air available to a material.

Aerobic. Living and functioning in the presence of free oxygen.

Aflatoxin. The mold toxin caused by *Aspergillus fluvus.* This is a coined word—"a" coming from the Aspergillus, "fla" from flavus, and toxin is added on—thus, aflatoxin.

Alkaline soils. Soils with a high pH, or one over 7. Alkalinity is usually the consequence of heavy sodium and potassium concentration.

Anaerobic. Living and functioning without air or oxygen.

Anion. An acid forming element. It is negatively charged and therefore migrates to the anode in an electrolyzing solution. It is the opposite of cation.

Annual. A plant that has a complete life cycle in one year.

Anthracite. A hard coal with fixed carbon between 86 and 98%. BTUs per pound run between 13,500 and 15,600.

Auxins. Plant hormones which govern the elongation and differentiation of newly formed cells. Auxins also regulate water movement in plant tissues, causing turgor changes. Turgor refers to the fullness of vessels and capillaries. Auxin-like hormones given off by roots during early germination are important in regulating competing weeds and seedzone bacterial systems.

Bacteria. Microorganisms, usually single cell, with shapes characterized as bacilli (straight or curved rods), cocci (spheres), or spirals. Bacteria are composed of protein and nucleic acid, and have chlorophyll, permitting a degree of photosynthesis. (See anaerobes and aerobes.)

Base saturation. The conceptualization on which standard soil tests are based. Base elements are calcium, magnesium, potassium and sodium. A soil should be saturated with usable cation nutrients up to 85 or 90% of capacity. This leaves 10 to 15% for hydrogen. According to the soil's capacity, calcium should occupy 60 to 70% of available sites; magnesium, 10 to 20%; potassium, 2 to 5%; sodium, .5 to 3%; other bases being variable.

Biennial. Plant that produces vegetative growth during the first year of growing season. After a period of storage or overwintering out of doors, flowers, fruits and seeds are produced during the second year; the the plant dies after the second year.

Biodynamics. A school of agriculture founded by the Swiss, Rudolf Steiner. It gained a foothold in the U.S. under the auspices of Ehrenfried Pfeiffer, and is practiced by many Americans as a valid form of eco-agriculture.

Biological transmutation. A theory authored by Louis Kervran of France that explains the constant change of one element into another.

Bituminous. A soft coal with fixed carbon between 69 and 86% for low to medium volatile, with BTUs between 14,500 to 15,600 per pound; and fixed carbon between 46 and 69%, with BTUs 11,000 to 15,000 per pound. Subbituminous coal has fixed carbon between 46 and 60%, with BTUs 8,300 to 13,000 per pound.

Bone charcoal. A charcoal made by firing bones at 1,800 degrees Centigrade.

Carbamates. Technically, an ester of carbamic acid. In trade, it is characterized as a broad-spectrum poison that is highly toxic to bees and fish, among other things. It is said to be less toxic to man—less being judged on the basis of make-it-to-the-door toxicology. The long range effects in terms of cellular mischief have never been evaluated properly.

Carbohydrate. An organic compound in which hydrogen and carbon typically occur in a ratio of 2 to 1.

Carbon. Chemical symbol C, atomic weight 12.0111, normal valence 4. It is a non-metallic element. Carbon is present in all organic and in a few inorganic compounds.

Carbon, activated. A carbon obtained by destructive distillation of wood, animal bones and other carbonaceous materials. Activation is conferred by the agency of heat at 800 to 900 Centigrade in most cases. The end product has high absorption capacity for gases, vapors and collodial solids.

Carbon, biogenic. An element of carbon represented by half of 1% of the total. It is indestructable to any type of disintegration. When a unit of carbon is reduced by 99.5% to humus, the balance represents biogenic carbon. It is called the spark of life carbon.

Carbon dioxide. A gas compound that forms when 1 carbon combines with 2 oxygens. In the case of human beings, it leaves the body chiefly when air is exhaled from the lungs.

Carbon-oxygen compound. As used by Siegfried Marian, it means a very pure humic acid with the molecules of carbon-oxygen arranged in crystallites in the form of a woven mesh. It was his conceptualization that the presence of such a compound in charcoal was due to trees taking up soluble salts as part of their metabolism.

Catalyst. A chemical or biologically formed substance which speeds up the rate of chemical and energy exchange reactions without entering the end product or being changed itself. (See enzyme.)

Cation. Positively charged base elements, either alkali metals or alkaline earths, which migrate to the cathode in an electrolyzing solution. A cation is the opposite of anion.

Cation exchange capacity (CEC). The capacity of soil components which exhibit surface-active adsorbing properties to hold positively charged nutrients as expressed in milliequivalents (ME). 1 ME of total cation exchange capacity represents the colloidal energy necessary to adsorb and hold to the colloidal system in the top 7 inches of 1 acre the following: 400 pounds of calcium; or 780 pounds of potassium; or 20 pounds of exchangeable hydrogen, etc.

Cause. As generally used, it means precedence. Every phenomenon, act, whatever, preceding another could be considered a cause.

Cellulose. Insoluble polysaccharide that makes up the cell wall of plant cells.

Chelate. A chemical or a biologically synthesized compound in which a metallic ion is bound up within the chelating molecule.

Chlorinated hydrocarbons. A family of insecticides and herbicides, including DDT, chlordane, aldrin, dieldrin, endrin and methoxychlor, and thousands more. Many are long-lasting and efficient in killing weeds and insect pests. They are also capable of entering the food chain. Almost all have been linked with degenerative metabolic disease. According to many of the world's top scientists, they have no safe level and they have no tolerance level. In agriculture they are used by farmers who have not mastered the precepts of eco-agriculture.

Chromosome. A rod-like body in the nucleus of the plant cell. It carries the hereditary message.

Cloudy charcoal. A carbon refined by Siegfried Marian with a great reactive surface impregnated with a molecular film of a very potent vitamin.

Coal. Consists chiefly of amorphous elemental carbon with low percentages of hydrocarbons, complex organic compounds and some inorganic materials.

Diamond. An allotropic form of carbon crystallized isometrically. Its carbon atoms are covalently bound by single bonds, chiefly in an octahedral structure. This makes it extremely hard.

DNA (deoxyribonucleic acid). A nucleic acid found in all living cells. Geneticists believe that genes such as are in chromosomes, consist of complex organic molecules composed mainly of this acid. Genes differ from each other according to their molecular structure.

Electron. A negatively charged particle that orbits the atomic nucleus.

Eloptic energy. A vibration not part of the electromagnetic spectrum in the extra low frequency range. This term is associated with the ELF work of Galen Hieronymus. (See odic force, X-force and orgone energy).

Enzyme. In nature, a catalyst produced in living matter, each related to specific combinations of minerals. It is a specialized protein capable of aiding in bringing about chemical changes. Enzymes promote a reaction without entering into the reaction, undergoing change themselves, or being destroyed.

Fermentation. An anaerobic process wherein carbohydrates or derivatives are decomposed by living microorganisms.

Fission. The splitting of an atomic nucleus induced by bombardment with neutrons from another source. Released neutrons propagate the process.

Fungus. Plant-like organisms that do not produce chlorophyll. They derive their sustenance by parasitic attachment to living organisms or by decomposition of organic matter. Siegfried Marian's insight, expressed in this book, fleshes out the total fungus picture.

Fusion. An endothermic nuclear reaction without radioactive hazard. The nuclei of light atoms unite or fuse.

Glucose. A simple—or monosaccharide—sugar. It is the prime fuel substance for most organisms.

Halogen. An electronegative element. Fluorine, chlorine, bromine, iodine and astatine make up the group. Fluorine is the most active. The rest are listed in the order of activity.

Hexagonal carbon. A common carbon form with six carbon

atoms in a ring for its crystal structure.

Hormone. In nature, a biologically energized chemical substance produced in one part of a plant that induces a growth response in another part at very low concentrations. Hormones also have been made synthetically. The most common are 2, 4-D and 2, 4, 5-T, herbicides which literally cause certain plants to grow to exhaustion and death.

Humate. A salt form of humic acids. Writing in *Economic Botany*, Everette M. Burdick put it this way. "Humus is the organic matter of soils that has decayed sufficiently to have lost its identity with regards to its origin. The most important and biochemically active group of the many degradation products of soil organic materials is the alkali-soluble fraction commonly called the humic acids. The salts of these humic acids are known as the humates."

Humus. Amorphous black or dark brown organic matter in soil derived from decaying plant or animal remains. Soil organisms preside over its production. Raw humus is formed under pH 3 to 4.5 conditions. Mull humus is formed in less alkaline soils. Humus is the end product of decay systems involving organic matter.

Hybridization. Using seed parents of differing genetic makeups to engineer crop performance. The term hybrid technically means a cross between different species. It is the first generation seed of a cross produced by controlling the pollination and by combining two, three of four inbred lines; one inbred of a single cross with an open-pollinated variety; or two varieties or species, except open pollinated varieties of corn. Commercial use of the term hybrid is restricted to the first generation following hybridization.

Hydrocarbons. These are organic compounds consisting exclusively of the elements carbon and hydrogen. Hydrocarbons are derived generally from petroleum, coal tar and plant sources. Olefinic hydrocarbons are produced by several plants, including the rubber plant, quayule and several members of the Euphorbiacae family.

Ion. An atom or atoms with an electrical charge.

Isotope. Elemental substances having identical chemical properties, but differing atomic weights. Most of the common elements have been shown to consist of a mixture of two or more isotopes. Isotopes of any one element have the same number of extra-nuclear electrons, and the same nuclear charge, but have differing nuclear masses. The importance in soil science lies in use of radioisotopes. These are basic elements that emit radioactive particles that can be detected by electronic apparatus or film. Radioisotopes of carbon, iodine, cobalt, phosphorus and other nutrients are widely used in plant and animal research to study translocation of chemicals and how they react in the tissue systems.

Leonardite. A humate material mined in the U.S. north central and northwest states. Humates are "this side" of coal. They have a chelating capacity. The Leonardite type of deposit has revealed anti-toxic effects in the presence of toxic doses of Cu, Ni, Fe, Zn, Mo, F, ethanol, pehnol salicylic acid and formaldehyde in culture of *Aspergillus niger.*

Lignin. A phenylpropane polymer in wood closely associated with holocellulose, and believed to be a binder for the holocellulose fibers. It comprises 17 to 30% of wood.

Micron. A unit of microscopic measurement. It is equal to 1/1,000th part of a millimeter. The symbol for micron is now MU as decreed by physicists.

Microorganism. A microscopic life form. Bacteria, molds and fungi are included under this definition. Viruses are not.

Mold. See fungus.

Mycorrhiza. Fungi in close contact with and entering into plant roots growing in biotically active soil.

Neutral soils. Soils with a pH at 7.

Neutron. A particle within the nucleus of an atom which has no charge.

Nucleic acid. Organic acid composed of joined nucleotide complexes. The best known are deoxyribose nucleic acid (DNA) and ribose nucleic acid (RNA).

Odic force. A vibration not part of the electromagnetic spectrum in the extra low frequency range. This term is associated with the ELF work of Baron von Reichenbach. (See orgone energy, eloptic energy and X-force).

Open pollinated. Generally refers to corn that is pollinated naturally by wind and insects, and not engineered genetically as are hybrid seed varieties.

Organic. A branch of chemistry which takes in all carbon-based compounds with the exception of certain binary compounds such as carbon oxides, carbides, carbon disulfide, etc. Some 3,000,000 organic compounds have been named. The term organic is used in the vernacular to designate a form of agriculture based on the primacy of organic matter.

Organic matter. Generally crop trash turned under to give microorganisms food, chiefly carbohydrates. Microbes have access to clay catchpens where other nutrients are warehoused. They balance carbon to nitrogen and ultimately break crop stover down to humus. When the soil is too low in fertility to serve both microbes and plants at the same time, the plants suffer first.

Organophosphorus insecticides. Compounds with very high acute toxicity. These compounds are better known as organic phosphates, a happenstance that has brought problems to the old organic gardening movement. Included among the organophosphates are parathion, malathion, mevinphos, and too many more. Using stone age testing methods, scientists have concluded that some organophosphorus materials are extremely toxic (acute) to most living things, others less so. Few tests have ever been conducted to cover several generations of test animals. Questions regarding the genetic effect on human beings over several generations have never been answered. Indeed, they have never been asked in most testing circles. Parathion poisoning brings on a roster of acute ailments that can stagger the imagination, if not the person. Included are general weakness, headache, sweating, nausea and vomiting, salivation, miosis, dyspnea, difficulty in walking, diarrhea, muscular fasciculation, disturbance in speech and consciousness, abdominal pain, fever, bronchophyaryngeal secretion, increased blood pressure, loss of pupillary reflex, cramps and cyanosis, among other things.

Orgone energy. A vibration not part of the electromagnetic spectrum in the extra low frequency range. This term is associated with the ELF work of Wilhelm Reich. (See odic force, X-force and eloptic energy).

Osmosis. A process by which water diffuses through a semi-permeable membrane from a side with less concentration. Diffusion continues until particle concentration is the same on both sides.

Paragonal carbon. A carbon that supplies the living sytem with oxygen via aeration.

Photosynthesis. Nature's process for manufacturing carbohydrates out of carbon dioxide and water with the use of light energy and the green light trapping pigment called chlorophyll.

Polysaccharide. A carbohydrate which yields more than ten monosaccharides on hydrolysis.

Proton. Fundamental particles of the nuclei of all atoms. The proton carries a positive electric charge.

Scientific system. Most of what is generally called the "scientific system" is not science at all, but merely procedural. The procedural aspect calls for setting up experiments that eliminate other possibilities, or it deals with making instruments that enable the investigator to find what he is looking for. The backbone of the scientific system has to do with asking the right questions. A scientist can only ask the right questions after his life has absorbed the experiences that lead him to a vision of the Creator's handiwork, hence the right question.

Synthesis. The creation of a substance that duplicates a natural product or creates a new molecular structure not found in nature via one or more chemical reactions.

Tetragonal carbon. A carbon that insulates.

Toxic genetic chemistry. A term used to designate man-made molecules of poisons used by some farmers to rescue crops from insect, fungal and bacterial crop destroyers. The word genetic is inserted here because the most commonly used molecules offend the genesis of life at the cellular level in plant, animal and man. Properly managed soil systems produce crops that do not require rescue with toxic genetic chemicals. The term also includes herbicides, which presume to rescue crops from weed takeover in the wake of management mistakes.

Toxicity. The capacity of a substance to inflict damage to a living entity, impair the central nervous system, or deliver severe illness or death when ingested, inhaled or absorbed by the skin. Effects of toxicity may be teratogenic, mutagenic or carcinogenic. The usual system for measuring toxicity is the LD_{50} or LC_{50} count, meaning the amount of a poison per unit of body weight it takes to kill 50% of the test animals. LD_{50} and LC_{50} do not explain offenses to the gene pool that do not necessarily involve death, but do account for degenerative metabolic disease.

Trace elements. Elements such as manganese, iron, copper, zinc, boron, etc., which are found as mere traces in soils and are used sparingly by plants. They are of vital importance to the growth and functioning of plants, as well as to the soil biological systems.

Trigonal carbon. A carbon that contains a reserve of free energy potential and becomes the feeding ground for microorganisms, of which bacteria causing alcohol fermentation is first in line.

von Liebig's Law of the Minimum. It states that the limiting factor in plant growth is determined by the absence of any essential nutrient, or its availability as an inadequate supply.

X-force. A vibration not part of the electromagnetic spectrum in the extra low frequency range. This term is associated with the ELF work of the Britisher Eemon. (See orgone energy, eloptic energy and odic force).

INDEX